Structures in Space

Springer
London
Berlin
Heidelberg
New York
Barcelona
Hong Kong
Milan
Paris
Santa Clara
Singapore
Tokyo

E. K. HALTEMAN

Bernard Abrams and Michael Stecker

STRUCTURES IN SPACE

Hidden Secrets of the Deep Sky

THE STECKER FILES

With 46 Figures
plus 12 Colour Plates

All photographs by Michael Stecker

Cover illustration (background): Antares–Rho Ophiuchus complex (image filename 0028.jpg)

ISBN 1-85233-165-8 Springer-Verlag London Berlin Heidelberg

British Library Cataloguing in Publication Data
Abrams, Bernard
Structures in space: hidden secrets of the deep sky
1. Astronomy
I. Title II. Stecker, Michael
520
ISBN 1852331658

Library of Congress Cataloging-in-Publication Data
Abrams, Bernard, 1957–
Structures in space: hidden secrets of the deep sky/Bernard Abrams and Michael Stecker.
p. cm.
Includes index.
ISBN 1-85233-165-8 (alk. paper)
1. Astronomy. I. Stecker, Michael, 1943– . II. Title.
QB43.2.A27 1999 99-35687
523-dc21 CIP

Printed in Great Britain

Typeset by EXPO Holdings, Malaysia
Printed and bound at the University Press, Cambridge
58/3830-543210 Printed on acid-free paper SPIN 10679649

Acknowledgements

Bernard would like to place on record his gratitude to his brother, Mike, for awakening a lifelong interest in astronomy, the late Walter Pennell and the very active Ron Arbour for their inspirational images, plus all Cotswold AS colleagues (particularly John Fletcher and Andy Packer) for their continued friendship and boundless enthusiasm, not forgetting Ann, Steve, Neil, Martin and Richard for their nocturnal tolerance.

Michael would like to thank Dr William Liller (retired Professor and Chairman of the Astronomy Department at Harvard University) for his encouragement and help, Dr Miguel Roth (Director of the Carnegie Las Campanas Observatory) for allowing use of the observatory facilities, Robert Hirshson for his computer expertise, and finally the "Mount Pinos Rat Pack" (Bob and Janice Fera, Bill and Sally Fletcher, James Foster, Martin Germano, Tony and Daphne Hallas and Dr Kim Zussman) – teachers and friends all.

Both Bernard and Michael are grateful to John Watson and colleagues at Springer Verlag for their help and advice during the production of this book.

Contents

Introduction

The visible Universe is a little more than 10^{10} years old – the best estimate is 15 billion years. It contains approximately 10^{50} tonnes of matter. Studying the ways in which this material is arranged, and the processes which shape it over time, provide the basis for the oldest science in the world: astronomy. Structures in space which have resulted from aeons of cosmic evolution combine immensity and delicacy in a way which gives astronomical photographs a rare aesthetic appeal.

This book uses the artful astrophotography and electronic imaging of Michael Stecker to illustrate these structures, and to illuminate the descriptions and explanations from Bernard Abrams which surround them. Selections from the Stecker Files have been chosen for more detailed description, including matters of imaging technique, as we expect there will be active astrophotographers as well as armchair astronomers in our audience – both are most welcome. This selection was extremely difficult; all the images are worthy of such consideration while speaking eloquently for themselves.

We are assuming that many readers will be comfortable with matters such as the magnitude scale of brightness used by astronomers, the right ascension and declination system of celestial co-ordinates, and the various films and instrumentation available to amateurs for astronomical imaging. Even so, we have tried to keep such technicalities to a minimum, except when providing details of methods used – for example within captions for photos and the more detailed discussion of some images – where it is supplied on the basis that such information will be of particular interest to those with relevant background knowledge.

The examples listed are a personal selection from a vast catalogue of potential targets. Further information can be found in publications such as *Nortons*, *Sky Atlas/Catalogue 2000*, *Uranometria* and *Burnham's Celestial Handbook*. Many modern computer-controlled telescopes have data files containing many tens of thousands of objects in most of the commonly used catalogues, allowing observers to locate objects at the touch of a button and rendering obsolete the difficult but rewarding task of learning the sky's fainter star patterns and then "star-hopping" to a particular target. Such is progress, we are told, and it certainly makes observing easier in some respects.

Any attempt to cover the variety of objects in deep space needs some sort of logic to guide both the writer and reader, and here we have preferred the natural hierarchy which is found on progressing outwards from Earth's position near the centre of the Solar System. Although this can also be tied in with the cycle of stellar evolution to begin with, the voyage out from our own Milky Way Galaxy into inter-galactic space necessarily breaks the timeline. On balance we felt that galaxies should be dealt with last rather than first, not least because they present some of the most difficult and enigmatic targets for the amateur astronomer.

The Earth is one of nine major planets known to orbit the Sun, and these – together with other objects in the Solar System – are reviewed in Chapter 1 as a starting point for our voyage into what some astronomers call the "deep sky". In keeping with part of the title of this book, aspects of the science behind each of the structures in space form a significant part of the narrative, and although this introduction continues by setting the scene as far as the astronomical picture is concerned, we must assume that some of our readers will already have more than a basic knowledge of astronomy and, if this is the case, or if a refresher isn't on your menu, you may wish to skip forward to Chapter 1.

Beyond the Sun's gravitational influence we enter the realms of inter-stellar space, where giant clouds of gas and space-dust are waiting to be seeded before spawning the next generation of stars. These dark clouds will become bright emission nebulae when the new stars are born within them. Beyond a scale of the separation between stars we move on to the wider structure and contents of our own Galaxy. Our home, the Solar System, is situated out to one edge of a spiral arm within this average-sized galaxy. Its collection of 10^{11} stars plus numerous star clusters, gaseous nebulae and dark matter pose timelessly for astronomers to make some beautiful images, which just happen to contain a wealth of scientific information – revealing the life cycle of stars, for example.

The Milky Way is one of two leading members of a small collection of galaxies known as the Local Group, which has the Great Andromeda Spiral Galaxy as its leading light. This assembly hints at the larger-scale structures found in the Universe: clusters and superclusters of galaxies. Most of these objects are at such immense distances and involve numbers of such magnitude that comprehending their true enormity presents real difficulties for the powerful but limited human mind.

The Andromeda Galaxy, our closest full-size galactic neighbour and an object visible to the unaided eye at a dark site, is at such a distance that light from it takes over 2 million years to reach us. While science fiction writers love the light year (the distance which light, travelling at 186,000 miles per second, travels in one year; this is approximately equal to six million million miles or nine million million kilometres), astronomers approve of the parsec and use the astronomical unit, both of which arise naturally when the Earth and the Sun are used in baseline measurements of cosmic distance. The astronomical unit (1 au) is the average distance of the Earth from the Sun, close to 150,000,000 kilometres. One parsec is equal to just over three and a quarter light years, and there are 206,000 astronomical units in 1 parsec.

Even this scale is too small to measure distances within and between galaxies, where kiloparsecs (1 kpc = one thousand parsecs = 10^3 pc) and megaparsecs (1 Mpc = 1 million parsecs = 10^6 pc) are needed. At this point the difference between a mile and a kilometre is less significant and only the number of noughts really matters; as the size of the Universe starts to sink in, it becomes even more remarkable that what we see is probably the leftovers after most of the matter and antimatter created in the Big Bang annihilated itself, leaving a "small" amount of matter and counterbalancing energy.

So our journey starts with some background information on the Solar System (Chapter 1), moves out into the Milky Way (Chapters 2 to 4) and ends in the realms of the galaxies (Chapter 5). Some observing ideas are included in Chapter 6, and we conclude with a pen portrait of Michael Stecker in Chapter 7.

Each of the types of celestial object covered is illustrated by a number of black and white images, plus a smaller selection in colour with many more on the enclosed CD-ROM. The information provided for those readers who are – or want to become – amateur observers, is by design non-exhaustive, since there is a wide range of needs dependent on the current experience and expertise of the reader. The best source of guidance in this regard is available from other active amateurs in any thriving local or regional astronomical society. A rich source of paper information can be found at www.adsabs.harvard.edu/abstract_service.html (an astronomy research web site).

A key advance in recent decades concerns the method used by amateur astronomers to record permanent images of celestial objects. After sketching came astronomical photography, and now we have CCD cameras (CCD = charge coupled device). While electronic imaging using a CCD unit is still restricted to relatively small targets due to the physically small size of the "chips" available to amateurs at reasonable cost, the vastly improved efficiency and a powerful ability to enhance electronic images using computer software has led to a rapid advance in quality. While Michael Stecker's technique involves analogue imaging (film) and doesn't make use of front-end digital technology, he uses a system (explained in Chapter 6) which can be described as getting the best of both worlds.

One final word of caution: the views through any telescope will usually be very different from those in this book, due to the lack of integration in the human eye together with its sensitivity (or otherwise) to particular wavelengths of light, and the Moon or planets form a better starting point for the novice visual observer. However, there is no denying the lure of more distant objects, and on any journey into the deep sky there are many fascinating structures in space to discover. We hope you enjoy the ride.

Bernard Abrams,
Cheltenham, UK

Michael Stecker,
California, USA

1 The Solar System and Beyond

The term "Solar System" is used to describe the set of objects bound to the Sun by gravity, the weakest of the four forces of nature but an important one in the Universe at large. The Sun dominates our Solar System. With a mass equivalent to more than 333,000 Earths, a million planets the size of our own would fit inside the Sun's globe. It is a glowing sphere consisting of mostly hydrogen gas, with a surface temperature of over 5,000° C and a core temperature exceeding 10,000,000° C.

An ordinary dwarf star, the Sun has been shining – by virtue of nuclear fusion reactions at its core which convert hydrogen into helium – for more than 5,000,000,000 years and is only about half-way through its expected lifetime of relative stability. During this period, the outward pressure from nuclear reactions at the core is balanced by an inward pressure due to gravity from the Sun's own mass. This stability has made stars like the Sun a common sight in the Universe.

Long-term stability does not imply similar short-term behaviour, and there is mounting evidence to show that relatively small-scale changes in the Sun's radiation and particle flux have a significant impact on the Earth's global climate. During the period 1645–1715 a mini ice age occurred, and astronomers at the time noted that in the entire period the Sun was devoid of sunspots – normally these wax and wane over an 11-year cycle. A period of "global warming" which occurred between 400 BC and AD 400 couldn't possibly be due to transport or industrial emissions, and presumably has a similar explanation. Astronomers of the day, had they been able to observe the Sun's surface safely, would very likely have seen a very spotty face.

Much recent research, particularly that due to Dr John Butler at Armagh Observatory, has established that centuries of climate change are directly attributable to the Sun. Man-made influences are not needed to explain any observed patterns past or present, but as astronomy is not studied particularly widely this key finding has escaped most commentators.

Once the Sun's available nuclear fuel has been used up, our climate will change much more dramatically. The Sun will shrink for a while before swelling to become a large red giant star, our oceans will boil dry and all life on Earth will come to an end. Don't panic: 5,000,000,000 years is a long time away, a third of the lifetime of the Universe, and the human race will certainly be travelling to new planetary systems long before then.

Planet Earth is the third object from the Sun. It has one natural satellite, the Moon, which is thought to have been formed by a collision between the young Earth and a Mars-sized object. Our planet is unique in one respect – it definitely supports intelligent life. While the question of past or present life on Mars continues to support missions to the red planet (life there is unlikely now given the surface conditions) and radio astronomers listen to the cosmos for signals from space, there is life in abundance on and around the surface of Earth.

The Moon provides most budding astronomers with their first "other-worldly" view through a small telescope, with craters, mountains, valleys and rilles to excite the observer. However, in terms of viewing faint objects deeper in space, the presence of the Moon in the sky is a complete disaster. The bright sunlight reflected from it illuminates the sky and submerges fainter objects which astronomers are equally keen to see. Nevertheless it may prove to be an advantage in the near future, as an observatory established on the dark side of the Moon (it keeps just under half its surface permanently turned away from Earth) would have splendid views out into space, without the bothersome radio and light pollution that is so troublesome on Earth.

An interesting phenomenon occurs when the Moon's orbital motion causes it to occult, or pass across our line of sight to, a more distant object, such as a star or planet. Stecker's image of the **Moon** shows another phenomenon. In addition to recording the occultation of the bright star Aldebaran on 10 April 1997, the use of a time exposure and careful tracking reveal the "night" side of the surface lit up by solar rays reflected from the Earth – known as earthshine. When the Moon is a young crescent, this phenomenon is known as "the old Moon in the young Moon's arms". Looking up from this part of the lunar surface at night, an astronaut would see a beautiful blue and white planet Earth in the sky, four times as large as the Moon appears to us, and something like one hundred times as bright (equivalent to five magnitudes on the astronomers' scale).

Of the remaining objects in the Solar System, the planets together with their satellites (Mercury and Venus have none) are generally the most conspicuous. These are grouped into three sets: the terrestrial planets (Mercury, Venus, Earth and Mars): the gas giants (Jupiter, Saturn, Uranus and Neptune); and finally Pluto in a class of its own (see Table 1). The inner group of four planets are small, dense, rocky objects like the Earth, while the gas giants are much larger and being composed mostly of gases lack any solid outer surface. The gas giants are also notable for their particulate ring systems, especially Saturn which is a telescopic favourite. Between these groups are the minor planets, also called asteroids. Michael Stecker's image of minor planet **Vesta** was taken on 18 September 1993 and is formed from a 45-minute exposure using a 10-inch f/4.5 reflector. It shows this third largest asteroid (510 km in diameter) passing between the much larger and more distant Helix Nebula – visually the largest planetary nebula in the sky – and Earth. Pluto, alone at the edge of the inner Solar System, is a very small icy object, which could represent a transition between planets and comets.

Table 1. Major planetary members of the Solar System

Planet	Mean distance from the Sun (km)	Equatorial diameter (km)	Orbital period (d or y)	Rotation period (d or h)	Mass (Earth = 1)
Mercury	5.8×10^7	4878	88 d	59 d	0.06
Venus	1.1×10^8	12103	225 d	243 d	0.82
Earth	1.5×10^8	12756	365 d	1 d	1.00
Mars	2.3×10^8	6794	687 d	1.05 d	0.11
Jupiter	7.8×10^8	142984	11.9 y	9.92 h	318
Saturn	1.4×10^9	120536	29.4 y	10.67 h	95
Uranus	2.9×10^9	51118	83.8 y	17.3 h	14.5
Neptune	4.5×10^9	49528	163.7 y	16.3 h	17.2
Pluto	5.9×10^9	2290	248 y	6.4 d	0.002

Beyond Pluto are a number of small icy objects in what is known as the Kuiper Belt, while beyond this lies the Oort Cloud, a second kind of comet nursery lying at the edge of the outer Solar System. If an object in either of these zones becomes disturbed in its orbit and falls towards the Sun, its icy constituents energize and begin to leak away into space, forming the tails of what we know as comets. Comets are basically dirty snowballs in space, consisting of frozen volatile materials (ices) such as water, ammonia and methane, together with dark silicate and organic matter. Historically they were seen as portents of doom ("when beggars die there are no comets seen, the heavens themselves blaze forth the death of Princes"); in reality they are no less interesting.

Comets represent some of the most primitive material in the Solar System, and there are suggestions that comet dust may have brought life to planet Earth. A bright comet is quite a rarity, although one or two really good ones should be seen in a lifetime. The Stecker Files contain images of three comets, two of which some readers may remember from the recent past: comets **Hale-Bopp** (1995 O1), **Hyakutake** (1996 B2) and **Tabur** (1996 Q1).

As with asteroids, comets have their own motion against the apparent paths of the stars. Therefore, the telescope's mount was programmed to guide on the comet's head instead of the background stars. This was done with the aid of a second guiding telescope piggybacked on the main photographic telescope, and centred on the comet's coma (head). The use of a CCD autoguider (SBIG ST-4) on the guiding telescope allowed for accurate tracking of the comet's motion and a clearer image of its outer structure – the nucleus of a comet cannot be imaged in this way.

Hale-Bopp and Hyakutake generated a lot of interest during their apparitions. To capture the former, Stecker used his 130 mm f/6 refractor working at f/4.5 and Kodak 1000 film, in a 20-minute exposure on 10 April 1997 near

Red Rock Canyon, California. The solarized image of Comet Hyakutake was also obtained from a 20-minute exposure guiding on the comet, through a 200 mm f/4 reflector on hypersensitized Fujicolor HG 400 film. The contour effect is due to scanning the image and applying a solarizing filter within Adobe Photoshop. This type of view is similar to the false-colour contour pictures of Comet Halley which were sent back from the ESA space probe Giotto.

Comet Tabur was equally interesting to the astronomical community since it appeared to be related to Comet Liller (1988 A1). It is presumed that both comets originated from a single parent body which broke apart at the previous return some 2,900 years ago, Comet Tabur being the smaller of the two pieces. This is not the first comet to do this and it will not be the last – at each return a comet loses substantial amounts of ice matrix, a process which can ultimately destabilize the nucleus and, together with tidal and other non-gravitational forces, lead to fragmentation. Michael's image resulted from a 15-minute exposure using a 155 mm f/7 refractor and Kodak Pro 400 PPF film on 12 October 1996.

Throughout the Solar System there is a thin sprinkling of space-dust and a very rarefied plasma, mostly from the Sun in the form of the solar wind. This stream of energetic nuclear particles varies in flux from time to time. When at its most energetic, this is the cause of ghostly patterns of light in the night sky known as *aurora borealis* in the northern hemisphere, and *aurora australis* in the south.

The separation between the Sun and Pluto is 40 au, and if we take 1 au as equal to 150,000,000 km then this is a distance of 6,000,000,000 km. To put this vast distance into perspective, the distance from the Sun to the nearest star – Proxima Centauri – is approximately 1.3 pc (4.26 light years, or 270,000 au). To make this comparison clearer, or perhaps more opaque for those who are rigidly astronometric, the number of astronomical units in a light year just happens to be very close to the number of inches in a mile. So in a model of the Solar System where the distance between Sun and Earth is 1 inch, the Sun and Pluto are 40 inches apart, while the nearest star is situated just over four miles away.

The journey to Pluto – never mind the stars – may seem to span an unreachable distance, but cameras let us cheat the void and in any case we have already explored deeper into space than this. One of the later *Pioneer* spacecraft, launched in the early 1970s to visit the gas giants, is still sending us signals from its nuclear-powered transmitter (solar panels are useless in the outer Solar System). This signal is equivalent to ten cellphones but can be detected by the world's largest radio telescopes at the current distance of ten billion kilometres. Scientists are still making use of it as part of the calibration routine in a search for radio signals from extra-terrestrial life.

Well beyond the Kuiper Belt and the bulk of the Oort Cloud, at a distance of 100,000 au from the Sun, lies the edge of the Solar System. At this distance the gravitational influence of the Sun is weak, and it is visible as nothing more than a bright star. Inter-stellar space beckons, and the first galactic astronauts could pave the way for future generations by leaving an ion trail spelling out the message "You are now leaving the Solar System – please drive carefully."

Our Solar System is special but not unique. Clues to the existence of extra-solar planetary systems orbiting nearby stars have been found from both terrestrial and satellite observations. The Infra-red Astronomical Satellite (IRAS) imaged a torus of cold proto-planetary material orbiting the star Beta Pictoris, which is situated along our line of sight towards the constellation Pictor at a distance of around 17 parsecs or 55 light years. The material is not hot enough to emit much light in the visible range, and can be detected more easily in the near infra-red (particularly by a telescope on a satellite above the Earth's atmosphere) where it is seen to stretch for some 60,000,000,000 km (6×10^{10} km) from the central star.

Another approach to locating the planets of other stars involves looking – indirectly – for massive, dark objects around some of the Sun's closest neighbours. These "invisible" planets are identifiable by the wobble they induce in the motion through space of their parent star, through the force of gravity. Some of the most interesting of these happen to involve stars which are of a similar type to the Sun, which astronomers classify as type G2V. There are thousands of such stars, and to date several candidates for star-plus-giant-planet(s) have been found. While the mass of an Earth-like object could not easily be detected, the presence of a larger planet similar to Jupiter, orbiting a Sun-like star, could signal the existence of what astronomers call a Goldilocks planet: a place where, like Earth, the temperature is just right. Just right for life to evolve, that is.

One of the earliest candidates for an extra-solar planet was implicated by the star 51 Pegasi. Not surprisingly, this is a star in the constellation Pegasus, the winged horse. When the motions of this star were analysed in detail, the timing of its wobble implied a rotation period for the suspect planet of about 4 days. Bearing in mind that one complete orbit of Mercury – the closest planet to the Sun

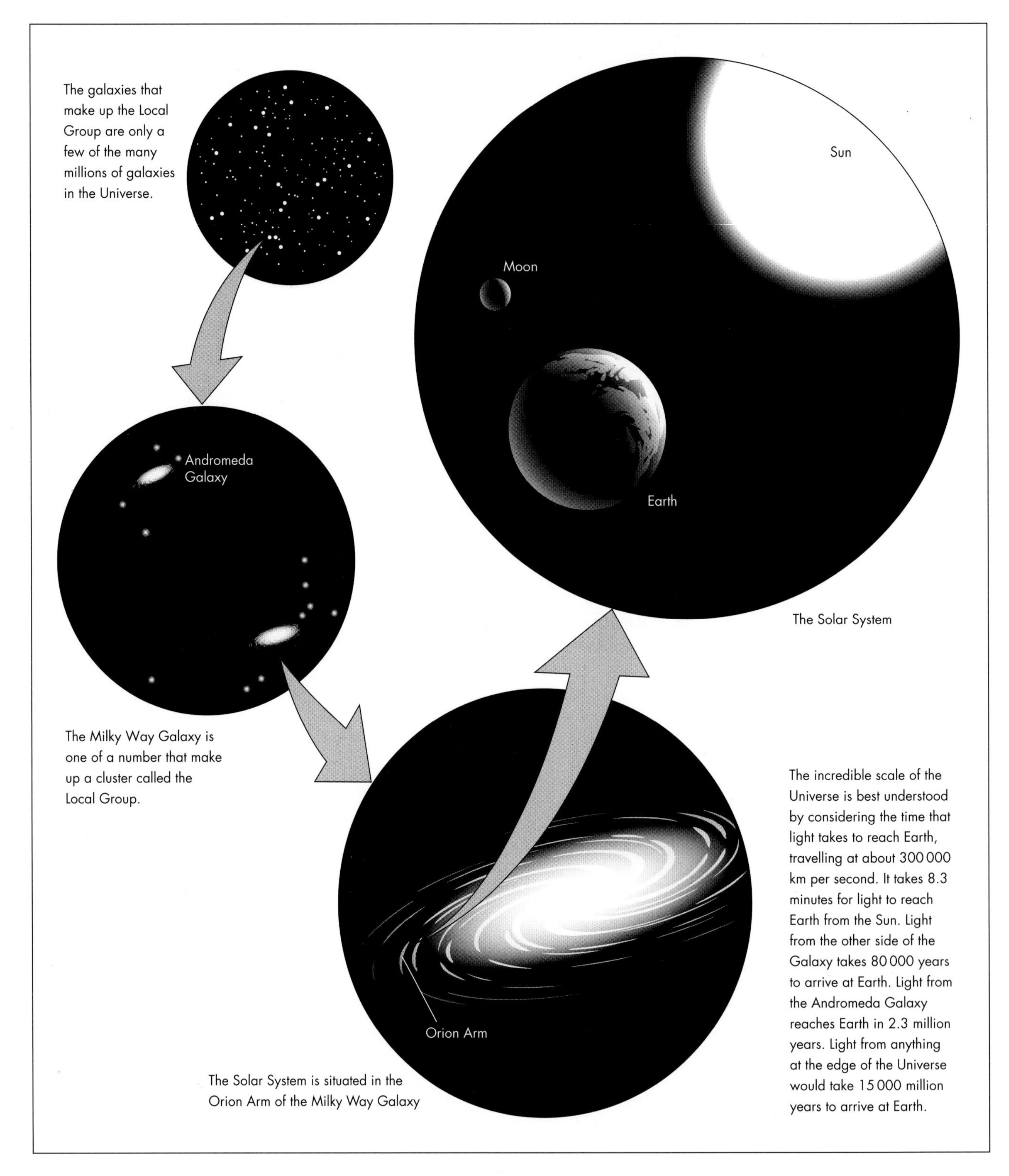

Figure 1 Our place in space.

Table 2. A selection of stars which have massive planets in orbit around them

Parent star	Planetary mass*	Orbital period (days)
51 Peg	0.5	4.2
47 U Maj	2.8	1060
70 Vir	6.6	116
55 Can	0.8	15
HD 114762	10	85

*Jupiter = 1

– takes 88 days, it is clear that the 51 Pegasi object must be orbiting extraordinarily fast and very close to the surface of its star. A selection of similar systems, which exhibit a wide range of mass and orbital parameters, is tabulated in Table 2.

What these observations demonstrate is that planets are not the rarity they were once thought to be. One model for the origin of the Solar System involved the gravitational pull from a passing star drawing a filament of material out of the Sun, which later cooled and condensed to form the planets. Given that the chance of a collision between stars, even when two galaxies hit each other head-on, is small, this type of process would make planets a rarity. A more popular idea is that the Sun and the planets formed together from a cloud of rotating and contracting gas and dust in space, known as the solar nebula. Observations of systems such as Beta Pictoris and 51 Pegasi make this scenario the more likely, and there are many such nebulae to be seen in external galaxies, but there are aspects of the formation of the Solar System which remain unresolved.

In terms of distance, the span of any such planetary system is a mere speck on the face of its galaxy, which in the case of our Milky Way has a maximum diameter of 30 kpc (100,000 light years, approaching 10^{18} km). Figure 1 shows the relationship between the Sun's family, the Milky Way and the Universe as a whole. Numbers in brackets refer to the chapters in which each of the structures is explored in more detail. Prepare to enter the deep sky.

2 The Nebulae

Dark nebulae, consisting mostly of cold molecular hydrogen, are revealed only by the presence of luminous material nearby or along the line of sight. Photographs of the Milky Way star clouds show numerous ghostly silhouettes, many adjacent to bright emission nebulae. Michael Stecker's wide-angle photograph of the **Milky Way**, viewed from our "insider" vantage point, should be compared with the image of external galaxy **NGC 891** discussed in Chapter 5, where similar dark nebulae and adjacent star fields plus bright nebulae along the galactic equator are seen edge-on from an "outsider" viewpoint. Pine trees on Mt Pinos are silhouetted against the Sagittarius Milky Way in this wide-field view obtained with a 3-minute exposure through a Pentax 165 mm f/2.8 lens and Pentax 6 × 7 format camera, using Kodak Pro 1000 PMZ-120 film.

Pioneering work in this area of wide-field astronomical photography was undertaken by Barnard, and a well-known catalogue of dark nebulae lists objects prefixed by the letter B. In earthly terms these clouds are very insubstantial, with densities similar to those found in a good partial vacuum, but the clouds are so vast that the total amount of material contained in an average dark nebula is sufficient to form many stars.

In keeping with the rest of the Universe, dark nebulae contain mostly hydrogen and helium, but there are other materials mixed in to form a cosmic soup, and an intoxicating one at that. Molecules in space are identifiable by their microwave signatures, and these signals – picked up by radio telescopes – have established the existence of alcohols such as ethanol and its more toxic sister-molecule, methanol. There are many dark nebulae which contain enough alcohol to sink a battleship, but collecting it would be a problem even though it would probably be "duty-free". While the conditions used to synthesize ethanol on Earth involve temperatures and pressures which result in 100 million million million molecules per cubic centimetre of gas, it is surprising to find such compounds being formed by chemical reactions in conditions where the mere 100 or so particles in the same volume rarely collide.

This material is also cold even by terrestrial standards, and certainly not hot enough to shine by emitting light in the visible region of the electromagnetic spectrum. This is a situation which doesn't last forever, as instability in the gas clouds can cause them to contract and spin up, eventually compressing the gas and space-dust enough in some places to form proto-stars. These are objects which glow by their own light, due to the heat of compression under gravity. Eventually, when the temperature at the core of these dense regions reaches about 15 million degrees Celsius, nuclear fusion reactions can begin, the outward pressure stops further collapse, and a star is born. Our Sun came into being in this way.

The instability in our own solar nebula, which led to the formation of the Solar System, might well have been caused by the death throes of a nearby star which ended its evolution as a supernova explosion. This fate awaits the most massive stars, and some binary stars (a pair of stars orbiting each other) as well. By seeding the rest of interstellar space with an evolved star's ashes, and sending a shock wave out into the cosmos, a supernova explosion can re-trigger star birth in its galaxy in what is effectively a cycle of stellar life. Such explosions represent one of the most energetic processes in the Universe. A single star undergoing a supernova explosion can outshine its parent galaxy – which contains hundreds of billions of stars – for many days. We take a longer look at supernovae in Chapter 3.

Our own Galaxy, the Milky Way, is long overdue for its next (visible) supernova. The most recent ones were seen in the years 1572 and 1604. A more famous example, which produced the Crab Nebula (a supernova remnant in the constellation Taurus) was visible in 1054. Unfortunately, we may have missed one or two supernovae since 1604. Our Solar System is located in a spiral arm of the Milky Way Galaxy, and these spiral arms are also home to giant molecular clouds and space-dust – the dark nebulae – which are very efficient at blocking out light from objects behind them. If a supernova were to occur behind a suitably large molecular cloud, we could not see it directly. Supernovae occur regularly in other galaxies, as will be seen in Stecker's image of the Whirlpool Galaxy.

The average extinction of visible light by space-dust grains in the plane of the Milky Way reduces the intensity of starlight by a factor of 2.5 for every kiloparsec. After travelling towards Earth for only 5 kpc, the light intensity of a supernova would be reduced to 1% of its original value. As the degree of attenuation varies widely from place to place, far greater extinctions than this are possible. Seeing a supernova is therefore a matter of luck, although if we happen to be too close it could be equally unlucky as the intensity of radiation from the explosion could reach dangerous levels.

If the material in a dark nebula was gathered up into lumps the size of golf balls, it would be far less efficient at blocking out starlight, and dark nebulae would not be as magnificent (or as awkward) as they are to astronomers. We must remember that – like all structures in space – they are three-dimensional, although we can only ever see them projected on to the plane of the sky as flat shapes. Their true size is immense, and one famous example in southern skies called the Coalsack stretches across many degrees of sky as viewed from our vantage point 200 parsecs away.

The grains of space-dust in dark nebulae are commonly graphite, a form of carbon, and silicate, containing silicon and oxygen atoms. A single silicate grain has a typical mass of 10^{-16} kg, and with upward of 10^{47} such grains a dark nebula would have a total mass equivalent to 10 Suns: the mass of the Sun is 2×10^{30} kg. However, there is almost always gaseous material associated with dust grains, mostly atoms of hydrogen gas, and as elemental hydrogen is typically ten times more abundant the actual mass of the nebula will be about ten times the value calculated for grains alone, in this case well over 100 solar masses.

Isolated regions of dark matter which can be seen inside bright nebulae, known as Bok globules, are believed to be the site of star formation. This process is one trigger which turns a dark (absorption) nebula into a bright (emission) nebula. With several such areas in a molecular cloud, dark nebulae are actually large stellar wombs capable of prodigious multiple births. Such globules can be seen in the photograph of the **Lagoon Nebula, M8**, and the extensive **IC 1396.**

IC 1396 (also known as Sharpless-131) covers nearly 6 square degrees of sky. For the positionally oriented amateur it is centred at celestial co-ordinates right ascension 21 hours 39 minutes and declination +57.5 minutes (epoch 2000). It was discovered in the late 19th century by E.E. Barnard with the 6-inch Cook refractor at Vanderbilt University, USA, and under today's sky conditions represents a challenging target for both visual observers and astrophotographers. Several conspicuous dark knots (globules) named by Barnard are visible in Stecker's image.

This is a relatively young nebula and its globules are thought to be only two to three million years old. It glows red (white in black and white photos) because its hydrogen clouds are excited by the central O6-class star, catalogued HD206267. IC 1396 actually represents the southern portion of a larger expanding molecular cloud called the Cepheus Bubble (10 degrees × 10 degrees, equivalent to a diameter of 120 pc). This "bubble" is thought to have originated from open cluster NGC 7160 to the north.

The brightest star in this area is Mu Cephei, which is situated in the north-eastern part of IC 1396. It is a semiregular variable star that varies in brightness by a factor of three (approximately). Mu Cephei was called the "garnet star" by William Herschel in recognition of the fact that it is one of the reddest naked-eye stars in the northern sky.

Obscuration of light emitted from behind a molecular cloud produces the characteristic silhouette of a dark nebula, some of which have intriguing shapes. A celestial shock wave can be seen in Stecker's wide field image of the **Horsehead Nebula (B33)** in Orion, a photograph which also shows the Flame Nebula below and to the left. In spite of the prominent view afforded by a time exposure – here, lasting for 45 minutes using a 200 mm f/4 reflector – the small angular size of this object and the faintness of the surrounding nebulosity make the photogenic Horsehead Nebula a very difficult visual target.

Other prominent dark nebulae with evocative names – usually more recognizably valid than the names of many of the constellations – include **B72** (the Snake) and **B79** (the Palm Tree). For both images Stecker used hypersensitized Kodak TP 2415 film, chosen for the excellent red sensitivity and fine grain which together produce high-quality wide-field images. Although quite insensitive to faint objects in its natural state, when this film is pretreated with a mixture of hot nitrogen and hydrogen – a process known as hypersensitizing – its response to low light levels is improved significantly without unduly compromising grain quality. A "fast" f/4 200 mm reflector was employed in both cases, using a single 45-minute unfiltered exposure.

However some of the best vistas are to be found in the interplay between dark and bright nebulae located in the same region of space. Such an arrangement is common, with a hot young star, or star cluster, forming in a molecular cloud and energizing some of the material nearby. Photographs of these structures give the appearance of continuous bands of light with gaps and rilles where the star density is less. We must remember that the star distribution in these lines of sight is more even than it seems, and that the presence of cold dark matter is blocking out starlight.

The images of stars on a long-exposure image blur into each other, and even small sections of the Milky Way can live up to their name even under such close scrutiny. Stecker's photographs of the **North America Nebula** and the **Pelican Nebula** also show the rich starfields to be found in and around the constellation of Cygnus the swan. While the naked eye will reveal similar objects in a dark moonless sky, the use of a red filter and a long time exposure (here, of 70 minutes duration) show their true glory, allowing even the casual observer to understand how each nebula got its nickname.

Often the light from glowing gas in space records on film or light sensitive electronic chips as a deep red colour. This is due to hydrogen in the nebula, as energized atoms of hydrogen radiate a pure, deep red light. To the eye, the same nebula would appear green in a large telescope. This difference arises because the human eye is not very sensitive to light at wavelengths equivalent to deep red, but very sensitive to the yellow-green colour emitted by oxygen in the nebula since this happens to correspond to that region of the spectrum in which the Sun emits most of its radiation. Evolution has taken care of the rest.

With photographic emulsions and electronic detectors the opposite is usually true, and glorious deep red images result with only a hint of yellow-greens. Explore some of the images on CD-ROM and this facet is immediately obvious. Another colour prominent in many images which combine dark and bright nebulae is the shade of blue characteristic of so-called reflection nebulae. This is a misnomer as the light is more scattered than reflected, but the name has stuck.

Since small particles scatter blue light much better than red (which, applied to our atmosphere, happens to be the reason why the sky appears blue and the Sun yellow), any dust grains near a bright star will scatter blue wavelengths towards us more efficiently. The result, regardless of the physics that brings it to us, is spectacular, as beautifully illustrated by the gaseous amnion surrounding some of the brighter stars in the **Pleiades (M45)** star cluster, clearly visible in both monochrome and colour images.

When looking at images of one of the best-known bright nebulae, the Great Nebula in Orion, it is difficult to appreciate that its true form is hidden from sight by a much larger yet unseen component. The dark material cutting into the nebula's central region (known to astronomers as the Fish Mouth) is part of a large molecular cloud which obscures much of the emission region and robs it of greater symmetry.

Perhaps the most famous pair of dark nebulae, which together typify the association with bright emission nebulae, include **B** (for Barnard) **33**, the Horsehead Nebula in Orion, and **B142/3** in the constellation of Aquila the eagle. Stecker's image of the Orion region around the Horsehead shows many dark nebulae inset in brighter objects, and was obtained in spite of the best efforts of Mt Pinotubo blocking the view with volcanic dust. B142 and 143 are side by side in the sky, ten degrees north of the celestial equator – B143 is the U-shaped one. The photograph of these dark nebulae was taken using an 8-inch f/4 reflector and a 40-minute exposure on hypersensitized TP 2415 film.

In addition, large regions of sky in the southern constellations of Scorpius and Ophiuchus are filled with star clouds and dark nebulae, making this part of the sky a favourite for both visual observers and astrophotographers. Michael Stecker's wide field image of the **Rho Ophiuchus Cloud** complex (one-hour exposure, red filter, hypered TP 2415 film, 300 mm f/2.8 lens) explains why in a manner which words could never convey.

For a more powerful illustration, go to the CD and examine the colour version, painstakingly constructed over two summer nights. It required three exposures, using a technique known as tri-colour photography (first developed by physicist James Clerk Maxwell in 1861) in which, as the name suggests, three filtered black and white images are combined to create an image in full colour. Here, three separate black and white exposures of the Rho Ophiuchus complex were taken with a Tamron 300 mm f/2.8 lens through red (42 minutes), green (80 minutes) and blue (90 minutes) colour filters.

The resulting three black and white negatives were scanned (digitized) so they could be processed in the computer with Adobe Photoshop software. Once in the computer the red, green and blue colours could be added together. After removal of an aeroplane trail in the green image, the three images were then digitally stacked – more details on this technique are given in Chapter 5. The result is a full colour image of the nebula and star field. Michael acknowledges the help of colleagues Bill Fletcher of Malibu, California, who first developed this technique and James Foster (Los Angeles) who modified it.

Many of the most subtle nebulae catalogued by Barnard are only revealed to best advantage by long-exposure wide-field photography as practised by Michael Stecker to bring you the illustrations used here. A dark site away from street lights is best, where a "rich-field" telescope or wide-angle camera lens will show the objects to best effect – see Chapter 5 for more observational guidelines. The southern hemisphere offers superb panoramas, some of which can also be seen from mid-northern latitudes. Binoculars may help if the sky is partly light polluted, but too much magnification will not help and a better solution would be to move to a darker site.

We have been referring to objects such as bright nebulae by using their designation in a catalogue drawn up by the French astronomer and comet hunter, Charles Messier. In this catalogue, the Orion Nebula is listed as **M42**. Another well-known sight to amateur astronomers, the Trifid Nebula (**M20**) is a bright nebula trisected by dark lanes which gives the nebula (and then the fictional plants) a name. One can then move on to objects such as M78 (Orion), and, more difficult, **IC 5146**. This image of the Cocoon Nebula in Cygnus is a composite of two 45-minute exposures by Michael Stecker using a 155 mm f/7 refractor, a medium-format Pentax 6 × 7 camera and Kodak Pro 400 PPF 120 film. NGC 7635 (the Bubble Nebula in Cassiopeia) has been glimpsed visually and is described as resembling a soap bubble in a steam cloud. NGC (New General Catalogue) and IC (Index Catalogue) are two lists of astronomical objects wider in scope than Messier's. There are several more – and more obscure – catalogues for the enthusiast to search through; indeed we quote from some of them in this book.

A very faint, challenging emission nebula situated only four degrees or eight moonwidths west of the famous planetary nebula M27 – the Dumbbell Nebula – is **NGC 6820** in Vulpecula. Stecker's image was obtained using a 10-inch f/4.5 Newtonian telescope with (25A) red filter and a two-hour exposure on hypersensitized TP 2415 film. Embedded in this emission nebula is a rod-shaped dark nebula reminiscent of the dark pillars characteristic of M16, the Eagle Nebula in Serpens. Scanning through decades of amateur astronomy magazines and several specialist books failed to locate another example of this rare nebula. Stecker's image appears to be virtually unique amongst published photos.

Another obscure but rewarding nebula is **Sharpless-205**, equally rarely photographed and listed in *Sky Catalogue 2000* with a brightness rating of 5, indicating one of the very faintest red nebulae listed. This is definitely one object that would be highly frustrating to photograph from light-polluted skies, as can be seen from the ghostly image which struggled to penetrate very dark Mt Pinos skies on the photograph obtained by Michael Stecker – courtesy of a 300 mm f/2.8 telephoto and a filtered one-hour exposure on hypered TP 2415 film. In order to increase contrast, a red (25A) filter was used with a 300 mm f/2.8 lens and hypersensitized Kodak Tech Pan black and white film. The filter suppressed background sky glow and allowed for a relatively long one-hour exposure at f/2.8.

A firm favourite of astrophotographers, the California Nebula (**NGC 1499**) in the constellation Perseus is easily recognizable as such in Michael Stecker's photo which resulted from a two-hour exposure through a narrow-pass red filter using a 200 mm f/4 reflector and hypered Kodak TP 2415 film. From Cheltenham, UK, such a long exposure would be impossible, the skyglow from light pollution burning out hypersensitized film after only 15 or 16 minutes at f/4. On Mt Pinos different rules apply, and the f-squared rule is extended by a factor of eight in this particular extract from the Stecker Files.

Dark clouds and bright nebulae form part of the interstellar medium (ISM) which is thought to be responsible for an important act of cosmic cookery. The elements lithium, beryllium and boron are quite common on Earth, yet they are much rarer than gold inside stars. This is because stellar nuclear reactions are very efficient at using up the nuclei of these elements. Yet there must be a reason for the higher abundance of these elements on our planet, and the ISM is most likely to blame. Collisions between cosmic rays and nuclei of elements such as nitrogen and oxygen can generate these three elements, so the lithium grease in our mechanical joints and the boron control rods in nuclear reactors are down to cosmic rays and the ISM.

External galaxies have similar dark clouds which we can observe from a vantage point many megaparsecs away in our own Galaxy, the Milky Way, as previously mentioned when comparing the view of Galaxy NGC 891 with our own Milky Way. These views are so similar because in the case of all spirals, dark nebulae are largely confined to the plane of the star-forming regions in the spiral arms. Spiral arms, rich in inter-stellar material, are also the site of open star clusters – structures in space that clearly remind us of that combination of immensity and delicacy that characterizes the deep sky. In Chapter 3 we will explore these, together with our Galaxy's pensioner star groups, the globular clusters.

Figure 2 The Milky Way.

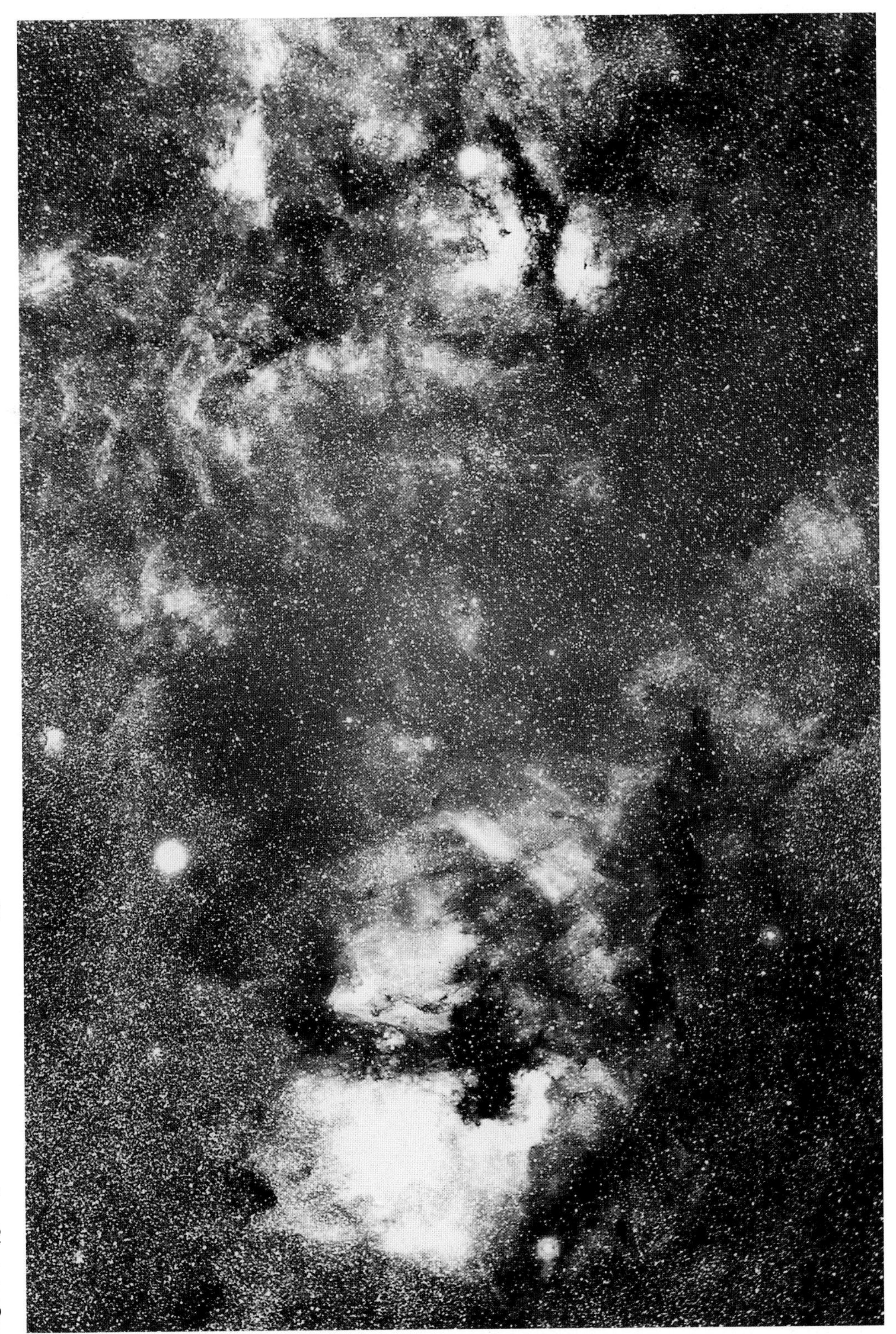

Figure 3 Cygnus (North America Nebula to Gamma Cygni).

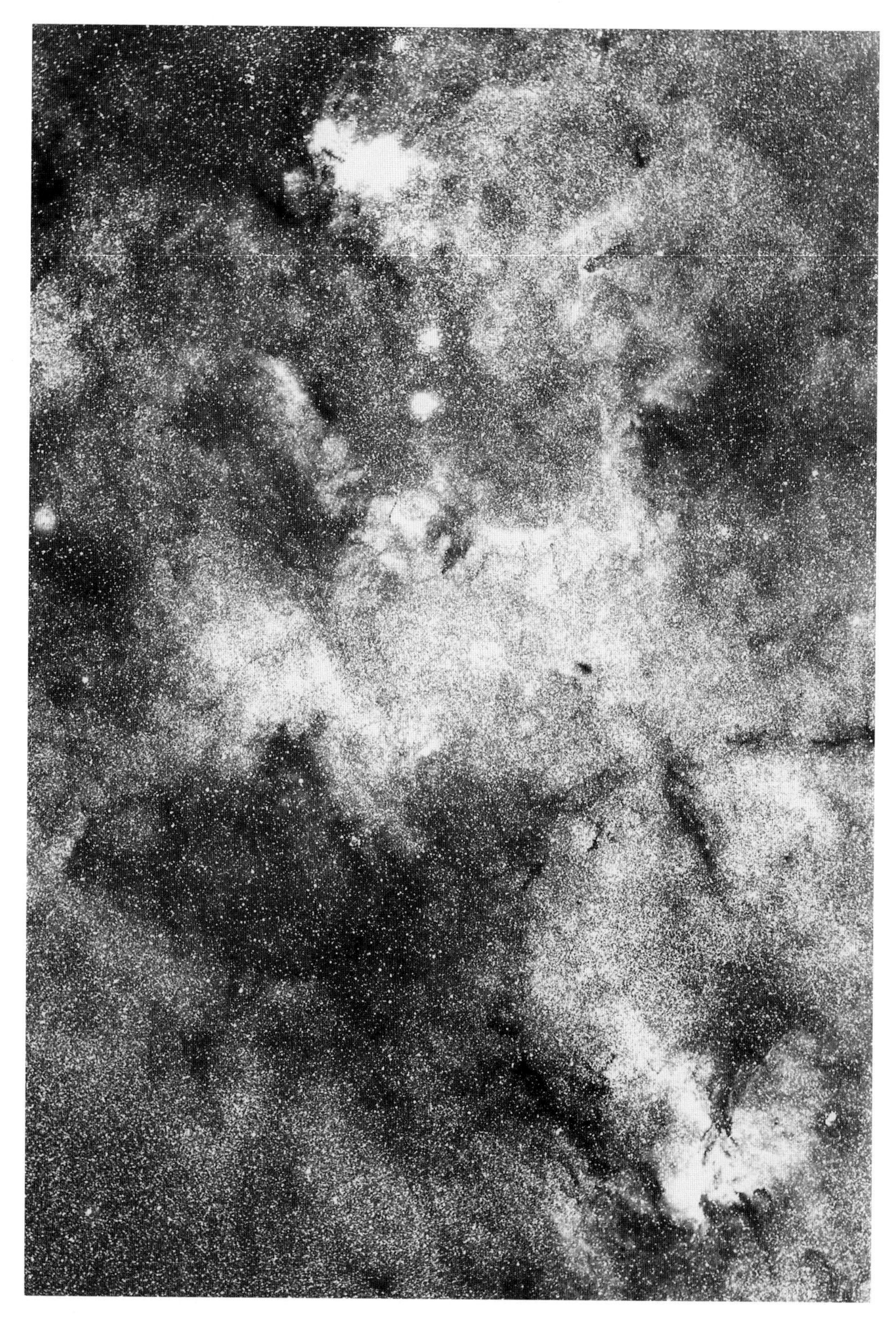

Figure 4 Milky Way (Ara to Scorpius).

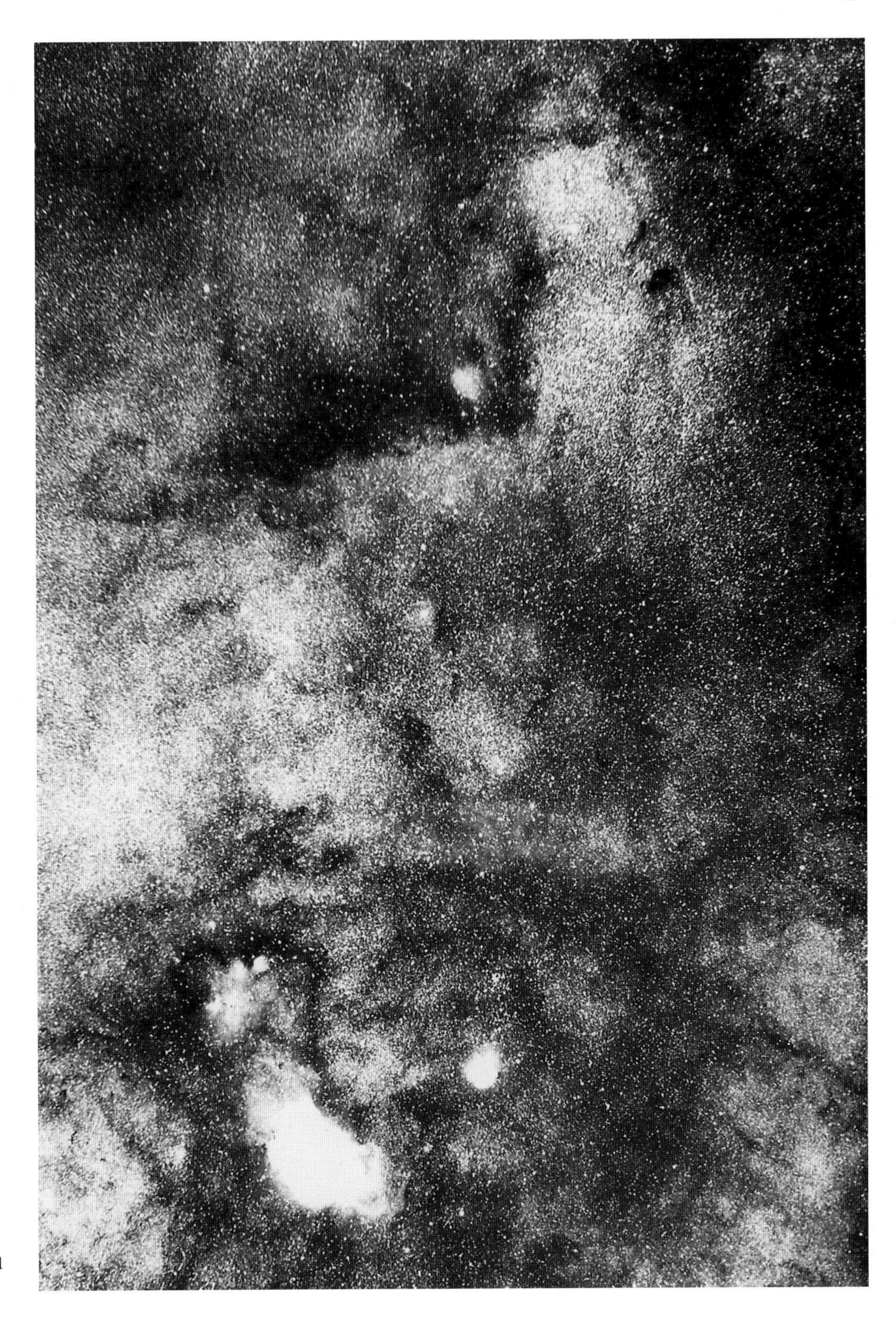

Figure 5 M8 to M24 and Sagittarius star clouds.

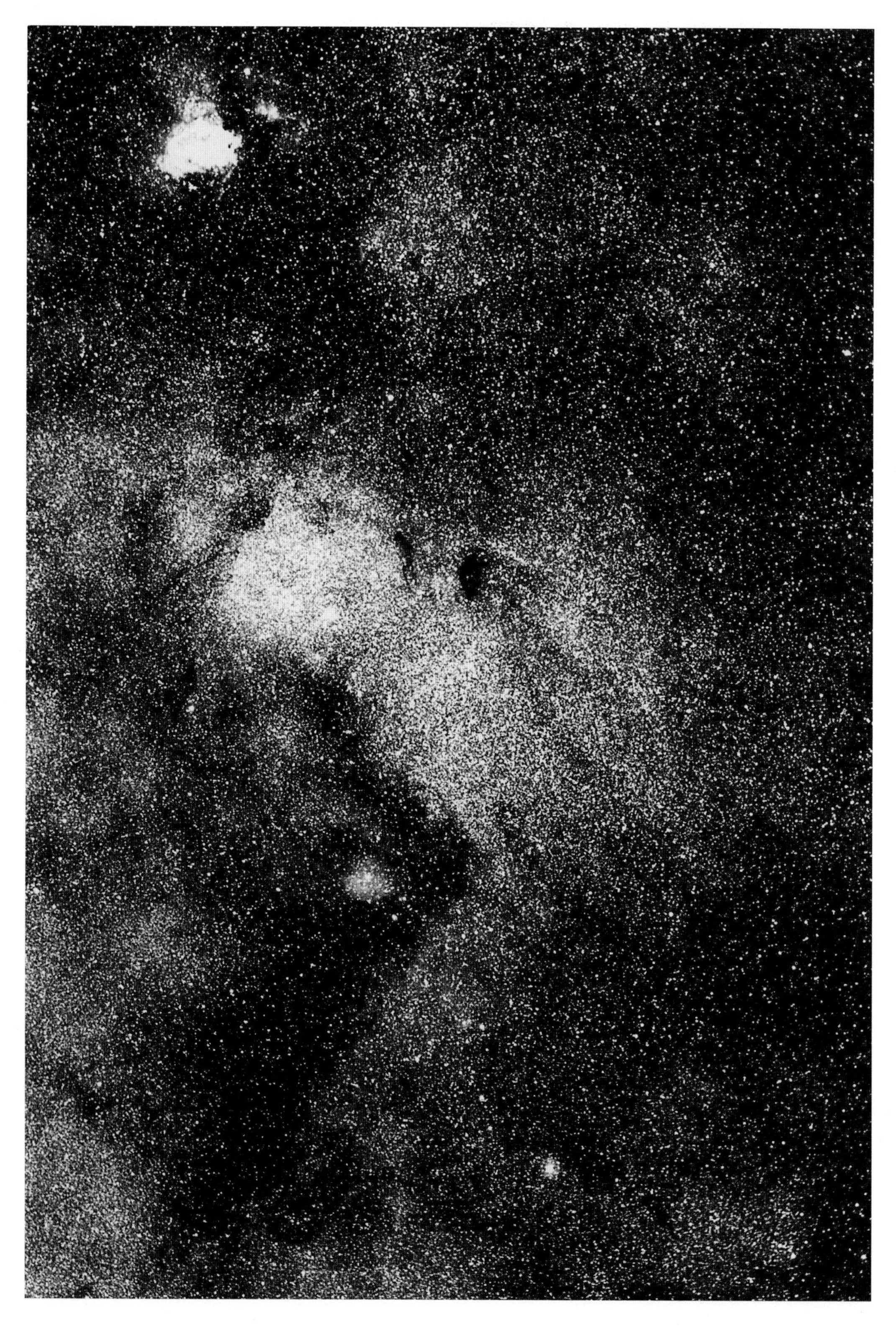

Figure 6 M24 to M17.

Figure 7 Cederblad-214 and NGC 7822.

Figure 8 Sharpless-205.

Figure 9 NGC 281.

Figure 10 Sharpless-142 (NGC 7380).

Figure 11 IC 1848 ("The Foetus").

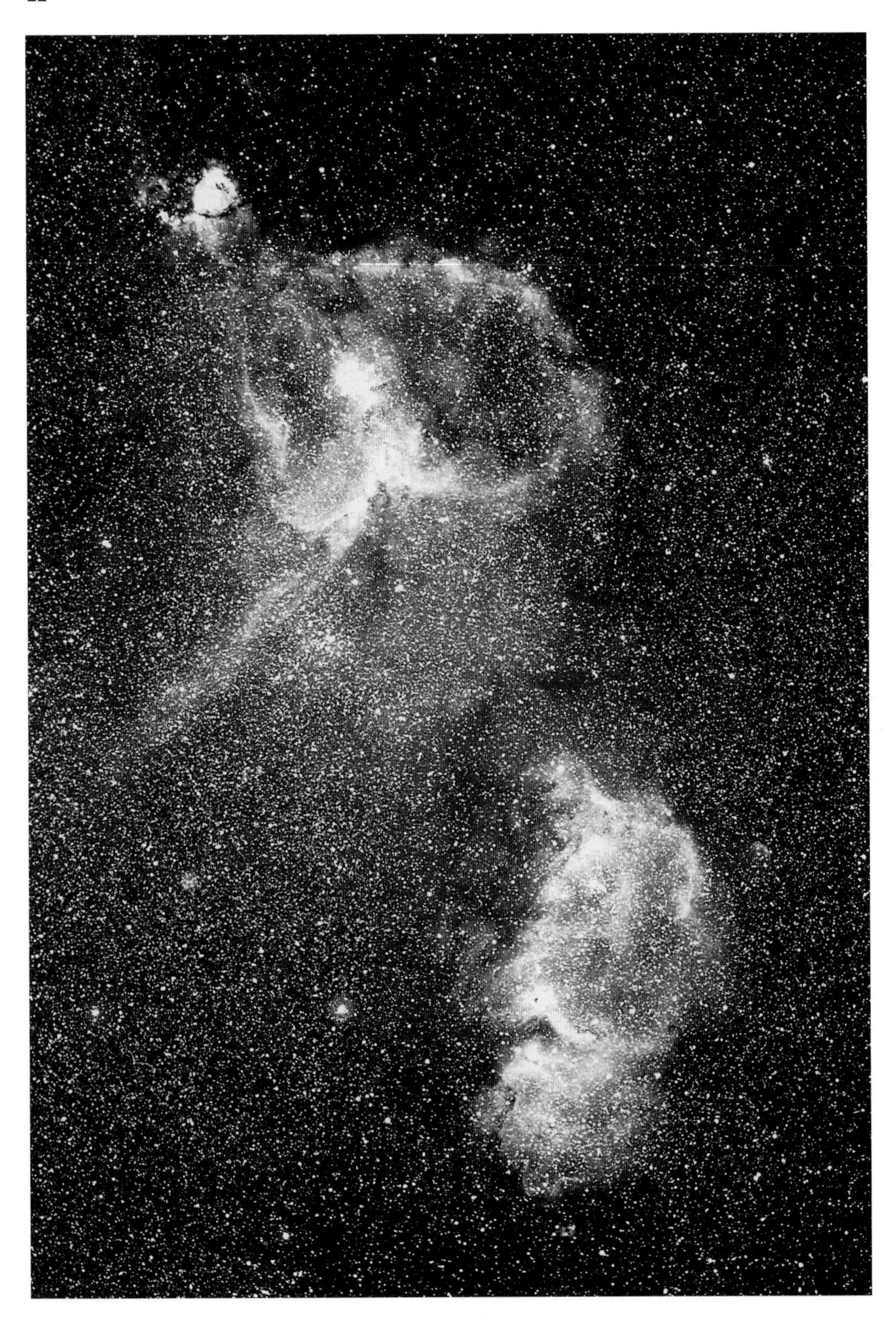

Figure 12 IC 1848 and IC 1805.

Figure 13 Rosette and Cone Nebulae.

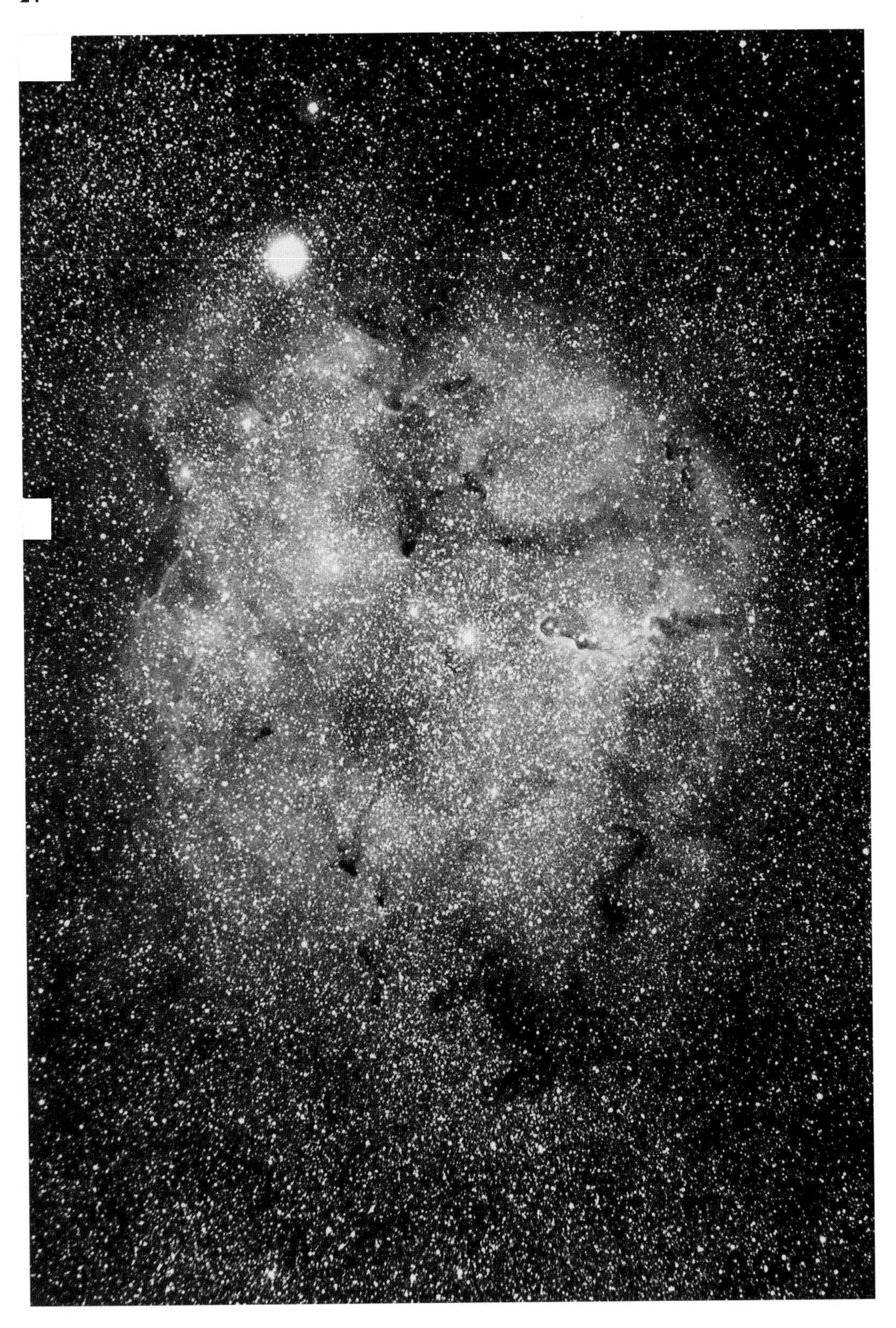

Figure 14 IC 1396.

Figure 15 California Nebula (NGC 1499).

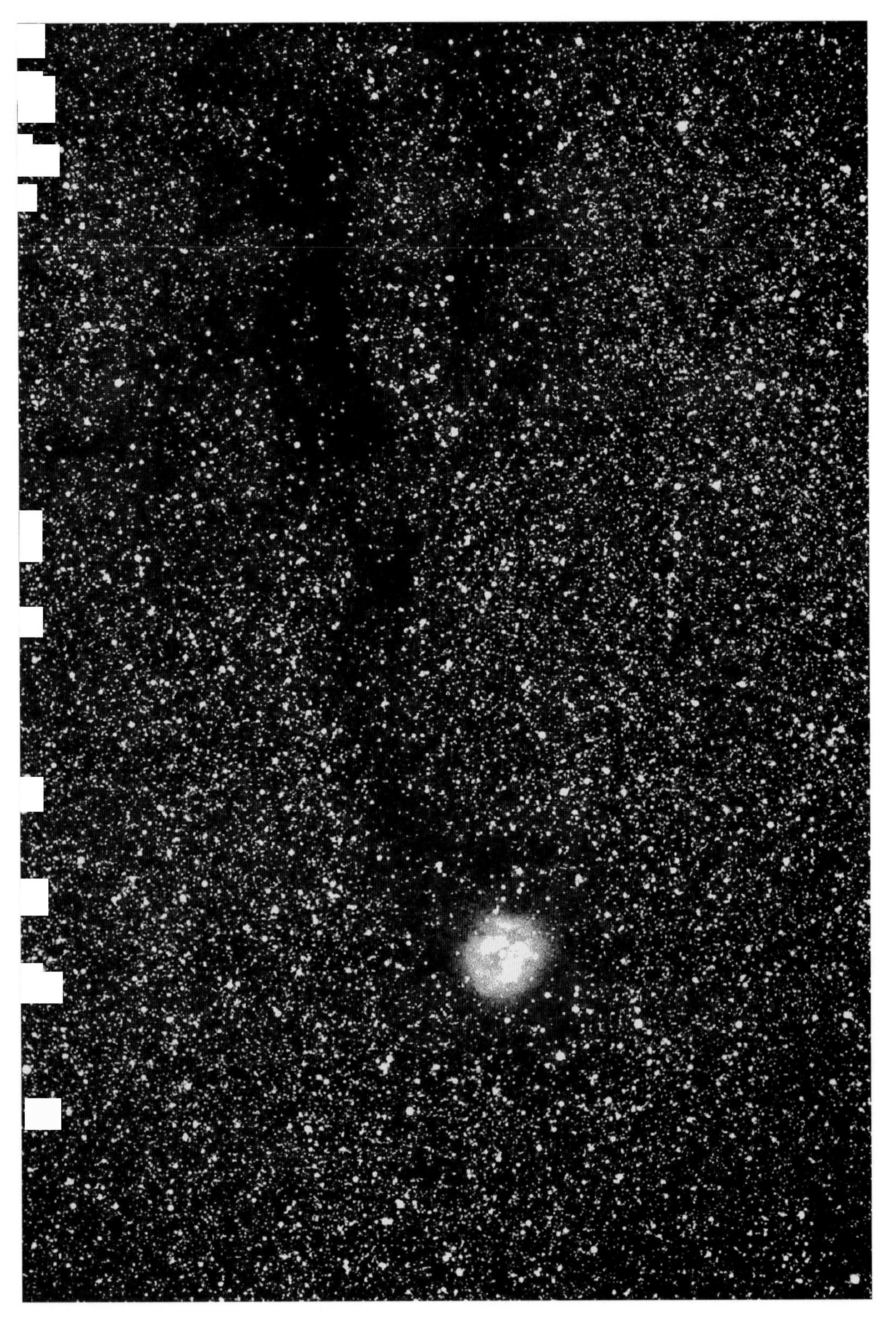

Figure 16 Cocoon Nebula (IC 5146). and B168.

Figure 17 North America (NGC 7000). and Pelican (IC 5067/70). Nebulae

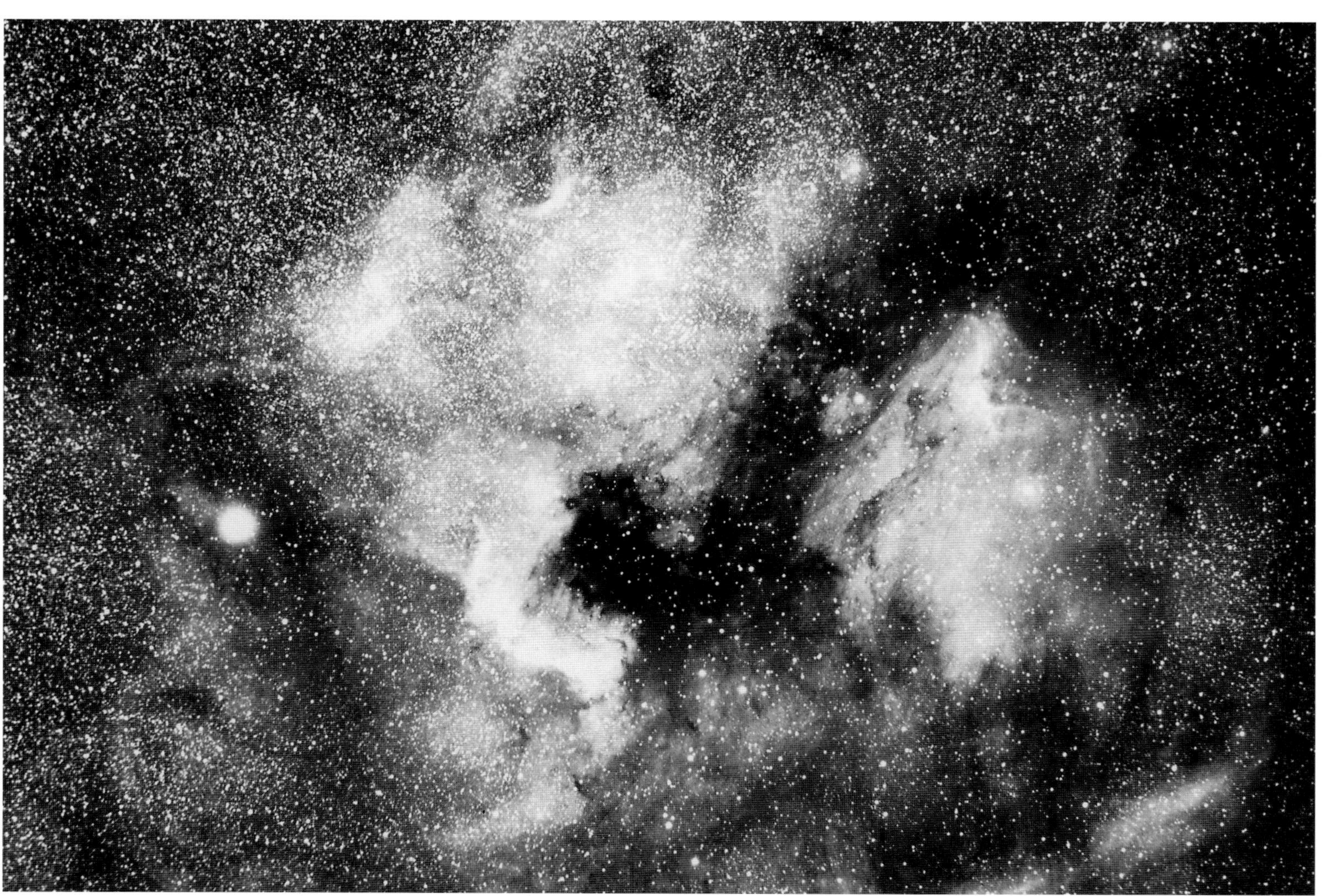

Figure 18 NGC 6820 in Vulpecula.

Figure 19 M8 (Lagoon Nebula).

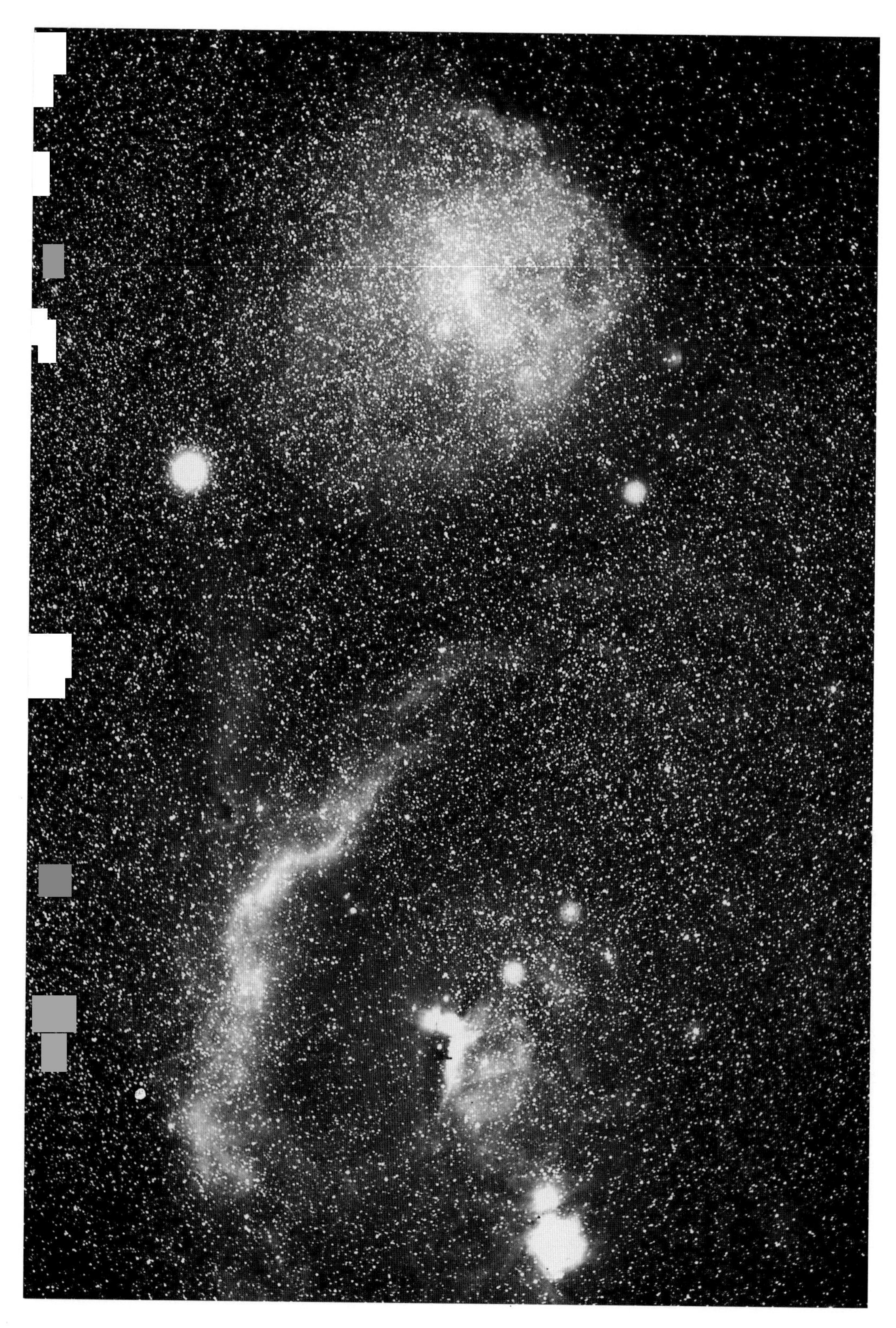

Figure 20 Sharpless-264, Barnard's Loop, Horsehead and M42 in Orion.

Figure 21 B33, the Horsehead Nebula.

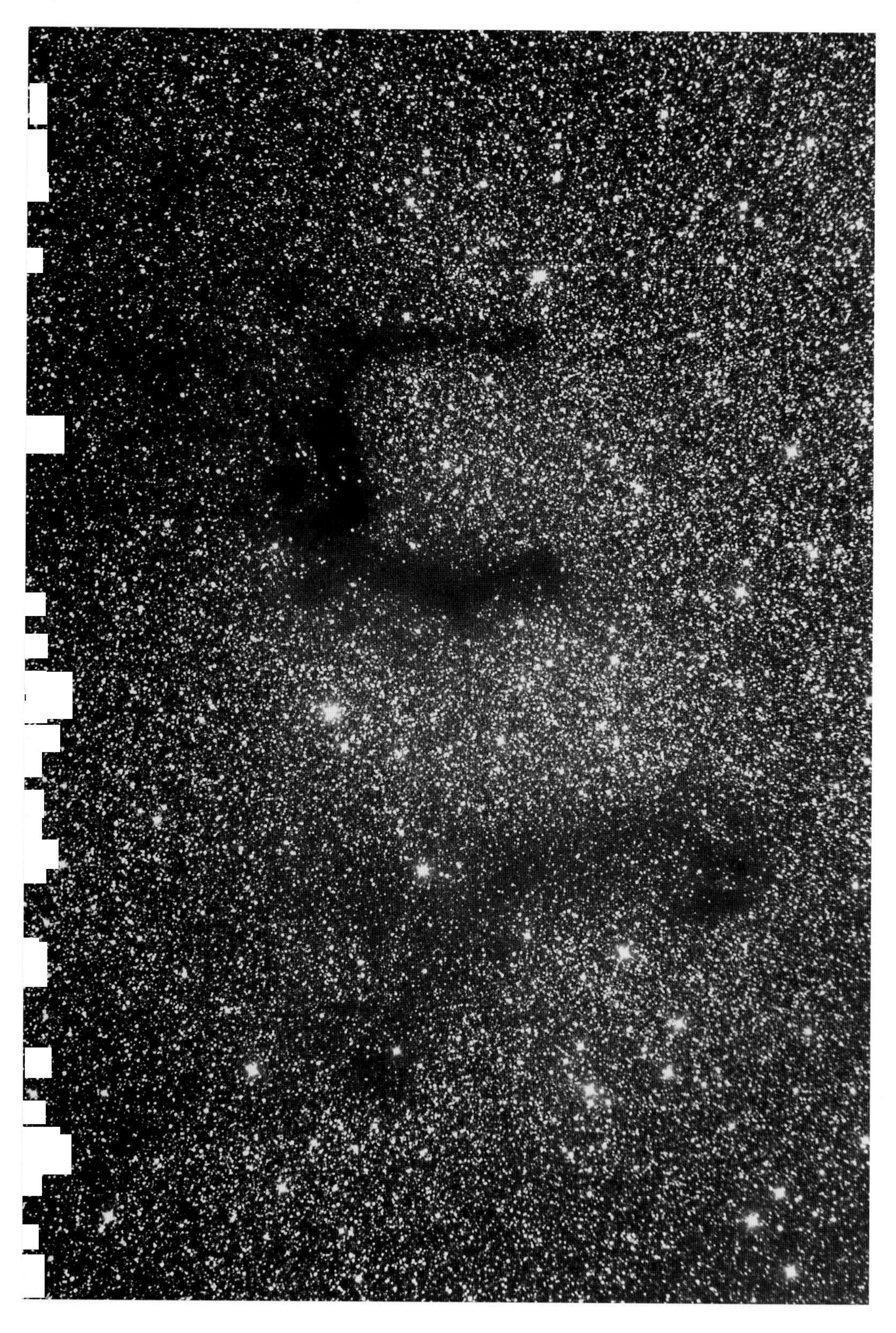

Figure 22 B142 and B143 in Aquila.

Figure 23 B79/B76 ("The Palm Tree").

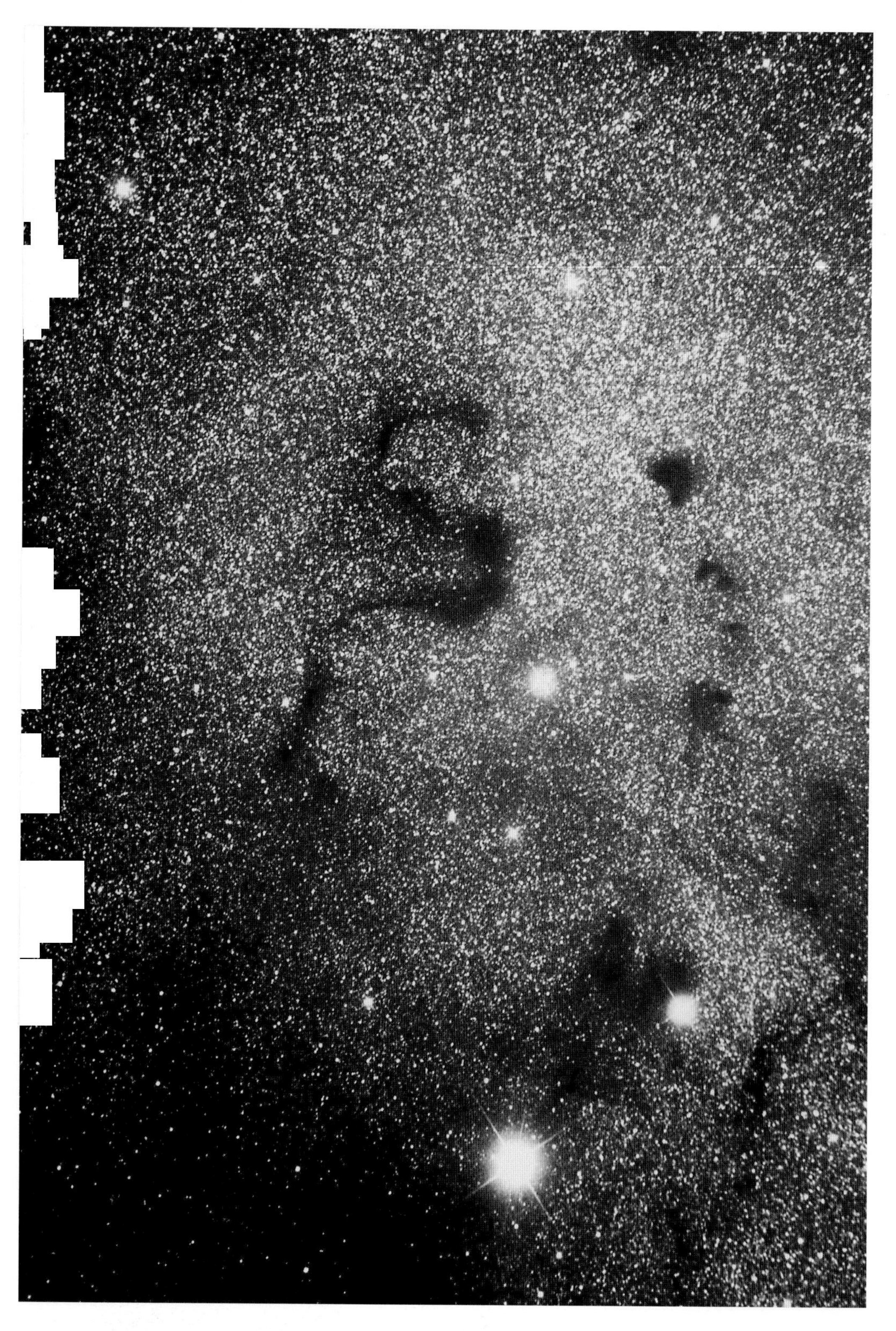

Figure 24 B72 ("The Snake").

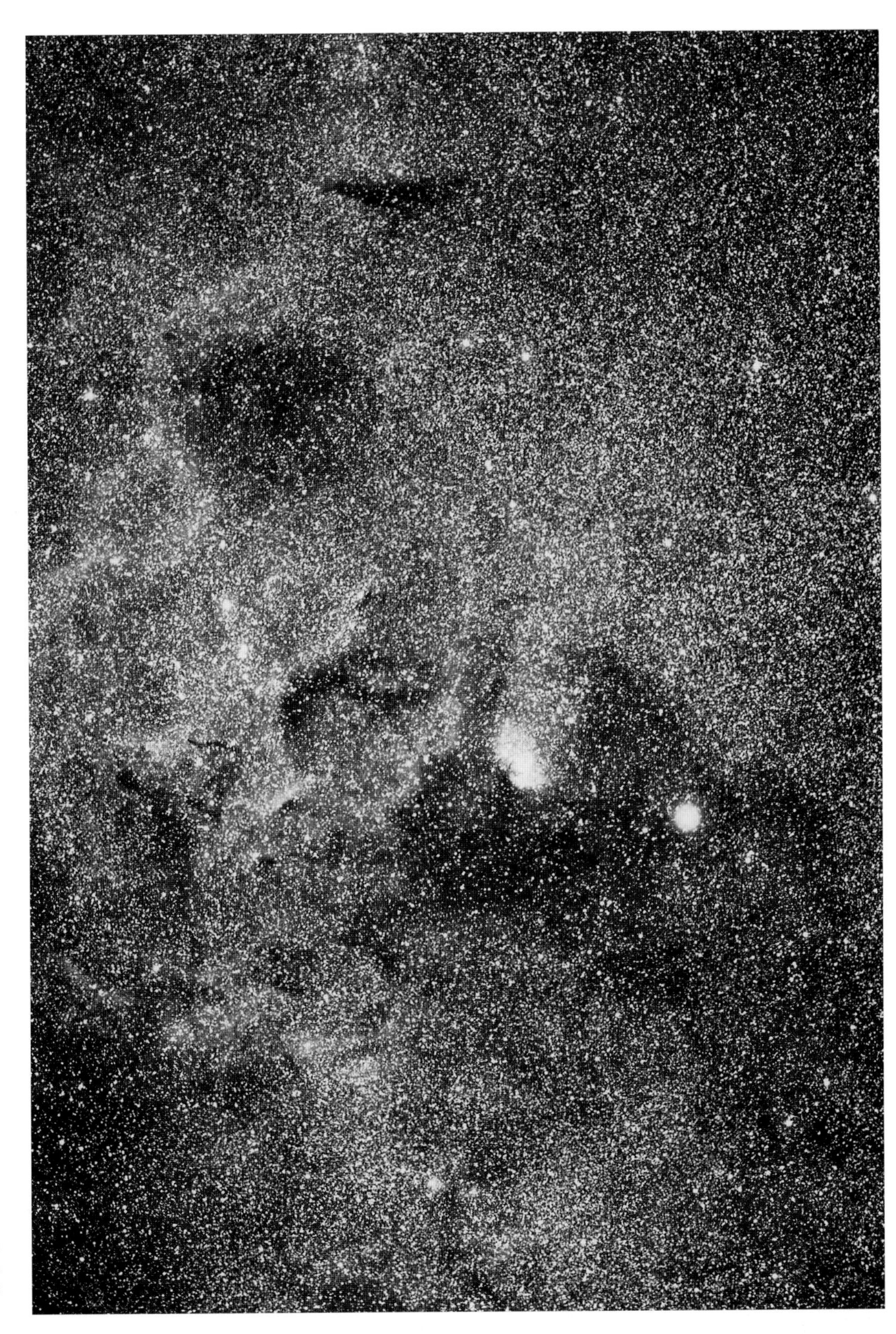

Figure 25 LDN857 ("Fish on the Platter"). with Sharpless-101 and Cygnus X-1

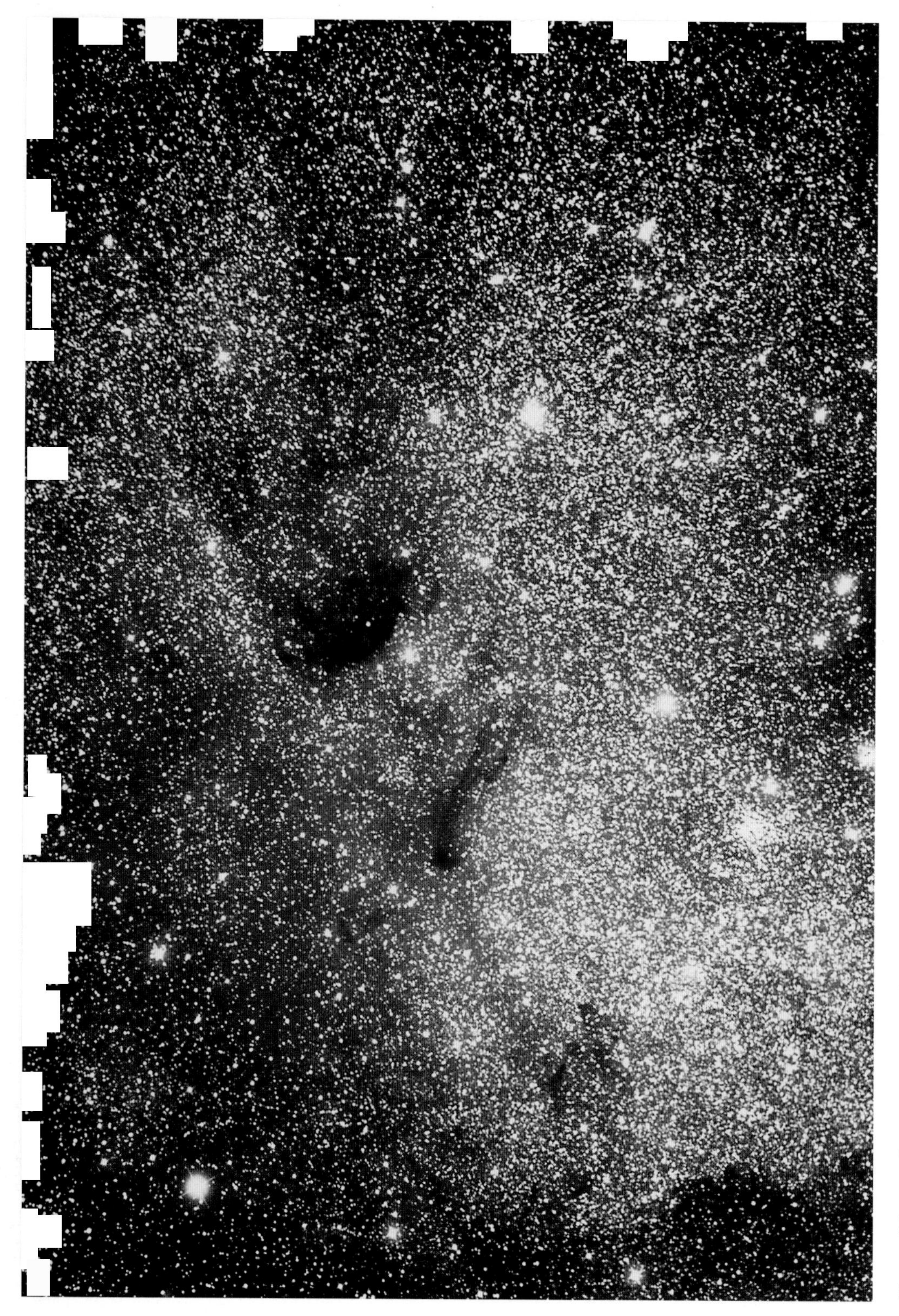

Figure 26 B92 and B93 in M24.

Figure 27 Cave Nebula (Sharpless-155).

3 Star Clusters

Only 15% of stars are formed alone inside their nebula nursery. The remainder are members of star clusters, of which there are two main types: open (or galactic) clusters and globular clusters. A third type, exemplified by OB associations, represents larger and looser collections, in this case of hot young stars. From our position in a spiral arm of the Milky Way, we can see many examples of both open and globular clusters. Although they share the same type of name – cluster – they could best be described as distant cousins.

In Messier's catalogue of nebulous objects, open clusters rank third in abundance, behind external galaxies and globular clusters. Many more can be found in the NGC and other astronomical catalogues. Open clusters are further categorized in terms of how tightly packed the stars appear to be. Although superseded by Trumpler's classification, the Shapley scheme is still used. Very loose, irregular groupings are given a "c" label, with more condensed, richer clusters ranging from "d" through to "g". The category "a" (asterism) can be used to designate very poor groupings which may not be true clusters.

Tombaugh-4 is about one-tenth of a degree in angular diameter, and is composed of about 40 stars. To illustrate the Trumpler classification, for Tombaugh-4 this is II 1 m, where the additional Arabic number indicates range in brightness and the letter for richness. In this case II means detached with weak concentration toward the centre, the 1 stands for a small range in brightness, while the m denotes a moderately rich cluster. Stecker's image shows this apparently diminutive object embedded in IC 1805. It was produced using a 155 mm f/7 refractor, Pentax 6 × 7 camera body and Kodak Pro PPF-120 film. Two colour images (50-minute single exposure time) were digitally stacked and a black-and-white print made from the magenta channel of the colour composite.

An open cluster is typically 5 pc in diameter and contains between 10 and 1,000 stars. While only hundreds of such objects are visible, there may be over 100,000 clusters in the Milky Way, mostly hidden from view by dark nebulae in the spiral arms. The majority are young in cosmic terms, ranging in age from 10 million to 1,000 million years. Consequently few highly evolved stars such as red giants are found in open clusters, as most member stars have formed relatively recently. Although they are born together and remain so for some time, gravitational forces from external objects and internal stellar encounters gradually disperse the member stars.

Very richly populated groups, such as the Wild Duck cluster (M11, in Scutum) and those situated out of the plane of the Galaxy (M67, in Cancer) survive the longest, as they are most able to resist the gravitational tug-of-war with the rest of the Milky Way. **Berkeley-17** is the oldest known open cluster at 12.6 billion years – close to the estimated age of the Universe – while **NGC 6791**, a very beautiful cluster in Lyra, is tied for second oldest cluster at 9.5 billion years. Both images were obtained using a 155 mm f/7 refractor on hypersensitized Kodak Pro 400 PPF film, NGC 6791 warranting five minutes extra over the 40-minute exposure used for Berkeley 17. Not far behind in the age stakes is the far northern cluster **NGC 188**, close to Polaris and estimated to be 7.2 billion years old. These ancient clusters provide astronomers with an insight into the progression between halo and disk formation in our Milky Way.

Careful observation and long exposure photography will reveal a relative lack of fainter stars in typical open clusters, due in part to the dynamical escape of low mass members. This process ejects stars from the cluster as a result of near-misses as stars orbit the centre of the cluster. Gravity acts through a slingshot mechanism to throw stars out into inter-stellar space.

The core of a cluster is usually all that we can detect visually. For example, the well-known Pleiades or Seven Sisters cluster (M45 in Taurus) is often listed as having an apparent diameter of one degree, equivalent to two full moons, yet member stars – located by astronomers using their common motion through space – have been found across 10 degrees of sky. For the Hyades cluster, also in Taurus, this scatter is over 30 degrees, equal to 60 full moons. At the opposite end of the image scale spectrum is a diminutive edge-on galaxy clearly visible in Stecker's image of M45 at the 3 o'clock position. This is designated in the *Uppsala Catalogue* as UGC 2838, but is curiously absent from *Sky Catalogue 2000* and *Uranometria 2000*. To capture it, two 45-minute exposures with a 155 mm f/7 refractor on 400-speed colour film were scanned and stacked digitally. UGC 2838 is an 18th magnitude galaxy, hardly a target for the faint hearted.

The mutual gravitational attraction of stars in a cluster might be thought to lead to an inevitable contraction, and eventual collapse, but this is not happening. Near the very centre of an open cluster, a hard binary may form early in its existence, involving two of the most massive stars, and this may be responsible for preventing continuous contraction. Interactions between this hard binary and other cluster members lead to it becoming tightly bound, simultaneously giving energy to other stars, spreading them out again and preventing collapse.

Some open clusters (such as M35 in Gemini) do indeed give the impression of having relatively vacuous centres with no sign of core collapse. A look at the photograph of **M7** (200 mm f/4 reflector, 40-minute exposure on hypered Kodak TP 2415 film) reveals a similar picture. Indeed, several clusters have been found to be stripped of a significant number of stars and are close to being totally dispersed.

Astronomers have suspected for some time that M45 is about 50 million years old, but developments in the theory of stellar structure hint that the age of this cluster may be underestimated by a factor of four. The reasoning behind the revised chronology concerns the evolution of stars. A star remains stable as a sphere (approximately) consisting mostly of hydrogen gas. In the core of the star, nuclear fusion reactions convert nuclei of atomic hydrogen into helium, and release energy in the process. It is this energy production which holds the star up for many millions – or even billions – of years, against a continuation of the gravitational collapse which formed the star from its parent nebula in the first place.

The period of stability varies according to the mass of the star, with low mass *dwarfs* taking longer than the lifetime of the Universe to become unstable. Massive *giant* stars, on the other hand, burn their nuclear fuel more quickly as their cores are much hotter. The most massive stars may exhaust their available hydrogen in only 10 million years or so, while dwarfs may shine steadily for well over 10 billion years. The Sun is thought to be near the middle of its stable phase, which is good news for the rest of the inner Solar System. Even so, the nuclear reactions in the Sun's core radiate energy into space which corresponds to a mass loss of four million tonnes per second. By solar standards, four million tonnes (1 tonne = 1,000 kg) is negligible.

Order-of-magnitude evolutionary timescales and stellar ages can be derived for stars and clusters of stars, but as mentioned above these may need revising, having been based traditionally on the assumption that only the hydrogen in a star's core is available for nuclear fusion. If a process known as *convective overshoot* takes place, the hydrogen-burning region extends well beyond the classical core, and this possibility could make the Pleiades up to 200 million years old.

Some areas of sky contain notable "clusters of clusters", such as the **Perseus Double Cluster (NGC 869/884)**, the Scutum area, and the Auriga trio (M36, M37 and **M38**, the latter seen here with the more distant **NGC 1907**). The black and white image of NGC 869/884 was taken with a 10-inch f/4.5 reflector, and a 25-minute exposure on hypersensitized Kodak TP 2415 film. The colour image on the CD shows a feature which is inevitably lost on the print, namely that there are a number of relatively bright evolved stars in the vicinity which add a touch of warmth to the otherwise cold blue tones of this beautiful object. Stecker's colour image was obtained using a 155 mm f/7 refractor and Kodak Pro 400 PPF-120 film (hypered) with a single 30-minute shot.

Analyses of such groups, where distance values make nearby objects truly close (and not a line-of-sight effect like the M38 pair) suggest that cluster formation may be initiated in more than one location by the same trigger event, possibly a nearby supernova explosion. Although the existence of a gravitational relationship within the Double Cluster is the subject of debate, the more dispersed Auriga clusters could similarly share the same origin, in the form of a cosmic shock wave, as they lie at similar distances and are therefore physically close in astronomical terms.

In addition to those mentioned above, some of the more attractive include M34 (Perseus) and M39 (Cygnus) – both good binocular targets – together with M41 (Canis Major), M52 and M103 (both Cassiopeia), and the powdery NGC 1245, easily found 3 degrees south-west of Alpha Persei. The aptly named **Owl Cluster (NGC 457)** is one of those rare objects which, with only a little imagination, has a name the origin of which can be understood. The cluster was photographed with an Astrophysics 155 mm f/7 refractor on hypersensitized Kodak Pro 400 PPF film exposed for 40 minutes.

Globular clusters are large, spheroidal groups of ageing stars, typically containing between 100,000 and 1,000,000 members, and with a diameter of around 50 pc. This is ten times the size of an open cluster, but since most globulars are also some ten times further away, they appear to be similar in size when viewed through a small telescope. One difference is immediately obvious to the visual observer, in that globular clusters are packed much more tightly together, a function of both reality and distance, so they are more difficult to resolve and individual stars cannot be seen as easily.

Unlike open clusters, which are young objects continually forming and dispersing in the Milky Way's spiral arms, globular clusters are amongst the oldest objects in the Galaxy, typically 10,000,000,000 years old. One or two notable globulars are visible to the naked eye, such as **M13** in Hercules (northern hemisphere) and **Omega Centauri** (southern). The image of M13 was taken from the southern California desert with a C-11 SCT at f/5.5 and a 30-minute exposure on Kodak Pro 400 PPF film, while Omega Cen used a 14-inch Celestron and represents the red channel of a colour image from hypered Fujicolor HG 400 film exposed for 45 minutes. Stecker's image of **47 Tucanae**, another magnificent naked-eye globular in the southern hemisphere, was taken using his 155 mm f/7 refractor with a 30-minute exposure on Fuji Super-G 800 film. These objects, without doubt the finest three globular clusters in the sky, are shown at their spectacular best in photographs such as these.

Other objects which stand out include M3 (in Canes Venatici), M5 (Serpens), Ms 10, 12 and 14 (in Ophiuchus), M15 (Pegasus), M22 (Sagittarius), and M92 in Hercules. The "inter-galactic tramp" cluster NGC 2419 (in Lynx) plus the Palomar clusters will together provide many hours of entertainment for the aficionado.

A noteworthy feature of our Galaxy's globular clusters concerns their distribution in a roughly spherical arrangement around the galactic centre. There are relatively few other stars (the so-called high-velocity stars) in the halo, with the vast majority of the Galaxy's stellar members located in the spiral arms of the galactic disc. Halo globular clusters could be a consequence of the more spherical shape adopted by the Galaxy in the first stages of collapse and star formation in the early Universe.

Globular clusters are relatively stable in comparison to their open cluster cousins. Observations of globular clusters show that, in spite of the obvious similarities, each object has distinguishing features. They contain mostly old, highly evolved stars such as red giants, in contrast to the hot young blue stars of open clusters. While open clusters are situated in or close to the plane of a spiral galaxy, globular clusters tend to be distributed in a spherical halo around their parent galaxy.

Distance has an effect on the angular size and resolution of globular clusters viewed from Earth, with the degree of condensation or compression inherent to the cluster also relevant to the view obtained. Catalogues and charts often classify globulars on the Trumpler scale. Clusters labelled I are most compressed and concentrated and are unlikely to show many individual stars, while those in category XII are loosely compressed and easier to resolve.

The problem of cluster core collapse, mentioned earlier in the context of open clusters, is even more relevant to the densely packed globulars. With this greater star density there is a distinct possibility that the core will quickly collapse, but once again it turns out that the formation of a "hard" binary star system in the core region can save the cluster from catastrophic meltdown. Indeed X-ray sources have been detected at the cores of some globular clusters, presumably those that have undergone core collapse to some degree.

Pre-collapse, the distribution of stars is more even across the cluster, indicating a type of globular known to astronomers as a King–Michie cluster. For those clusters that have undergone a recent collapse halted by hard binary binding energy – the Henon clusters – there is a very small core. Examples of King–Michie clusters include M5 and the Great Globular Cluster in Hercules, M13. Henon clusters include M15 in Pegasus. The images of **M4** and **NGC 288** shown here also happen to provide views of other structures in space: a cloud complex and a galaxy respectively. The print showing cluster NGC 288 and galaxy NGC 253 is the result of a 45-minute exposure using a 200 mm f/4 reflector and hypersensitized Kodak TP 2415 film.

There are a number of puzzles involving globular clusters. One concerns the existence of "blue" globulars as found in the Large Magellanic Cloud (LMC), a satellite galaxy of the Milky Way. Morphologically the blue globulars resemble normal globular clusters: they are large, spherical, and heavily populated – but their visual appearance differs in that they have overall colours which are, as their name suggests, much bluer.

The differences between LMC "blues" and Milky Way "reds" could be due to a different chemical composition in their member stars. A lower abundance of heavy metallic elements would lead to lower opacity in the stellar atmosphere revealing hotter (and therefore bluer) regions within. Very few such objects have been studied in detail and are the subject of continuing controversy.

One similarity between globular and open clusters is their susceptibility to disruption by tidal forces. Sitting out of the plane of the Milky Way for much of the time, globular clusters generally have a more peaceful existence, but their orbits around the galactic centre will, from time to time, take them through the galactic disk. At such stages there is a sudden rise in the strength of gravitational forces from outside the cluster, which will strip it of its outer layer of stars.

The entire stellar population of the Milky Way's halo is thought to consist of stars torn from globular clusters, and possibly one or two dwarf satellite galaxies which dissipated long ago. At present there are between 100 and 200 globular clusters, but in the distant past there may have been twice as many. Since our Galaxy formed, many of those globular clusters originally witnessing the Milky Way's adolescence will have been scattered around the halo.

Stars in an open cluster will be separated by distances not too dissimilar from those in our region of the Milky Way, with nearest neighbours just a few light years away. In a globular cluster core the population density is much greater, and although it is far less than appearances on photographs suggest, it may be sufficiently high to produce occasional collisions between stars, and certainly some close shaves. Under these circumstances stars may have their outer envelopes of gas stripped away, revealing a hotter (and bluer) inner layer. This process could account for the small number of blue stars found near the cores of some globulars, where only old red giants are expected, and could go some way – together with differences in heavy element abundance – to explain the existence of the LMC's blue clusters.

Open clusters are excellent targets for the binocular observer or small telescope owner. Many lie just beyond the naked-eye threshold and are revealed and resolved by the most modest optical aid. Lying within the star-forming regions of the Galaxy's spiral arms, in the galactic plane, open clusters are irregular collections of up to a thousand or so stars bound together gravitationally. It is difficult to avoid the feeling of awe which open clusters inspire when viewed with a large telescope at low power. There are scientifically useful avenues of work open to suitably equipped and dedicated observers of open clusters – monitoring flare stars, for example – but it is more often for aesthetic reasons that open clusters in the winter and summer Milky Way are observed.

Amateur astronomers intending to study globular clusters need to use large apertures, as recommended for any deep sky observing, but longer focal lengths (and/or higher magnification eyepieces) will be useful in resolving some of the more compressed objects. Autumn and spring skies provide the best collections, since the obscuring matter of the Milky Way is less of a problem; Hercules, Sagittarius and Ophiuchus offer many excellent targets which illustrate the range of morphology displayed by these objects.

Working with different apertures – or with a range of exposures in photography – can produce very different impressions of globular clusters, more so than other objects. Small apertures and short exposures reveal the distribution of luminous giants, while larger telescopes or moderate exposures will reach stars evolving towards giant status. Only the largest visual telescopes, and deepest exposures, will penetrate down to the faint dwarf stars in such clusters.

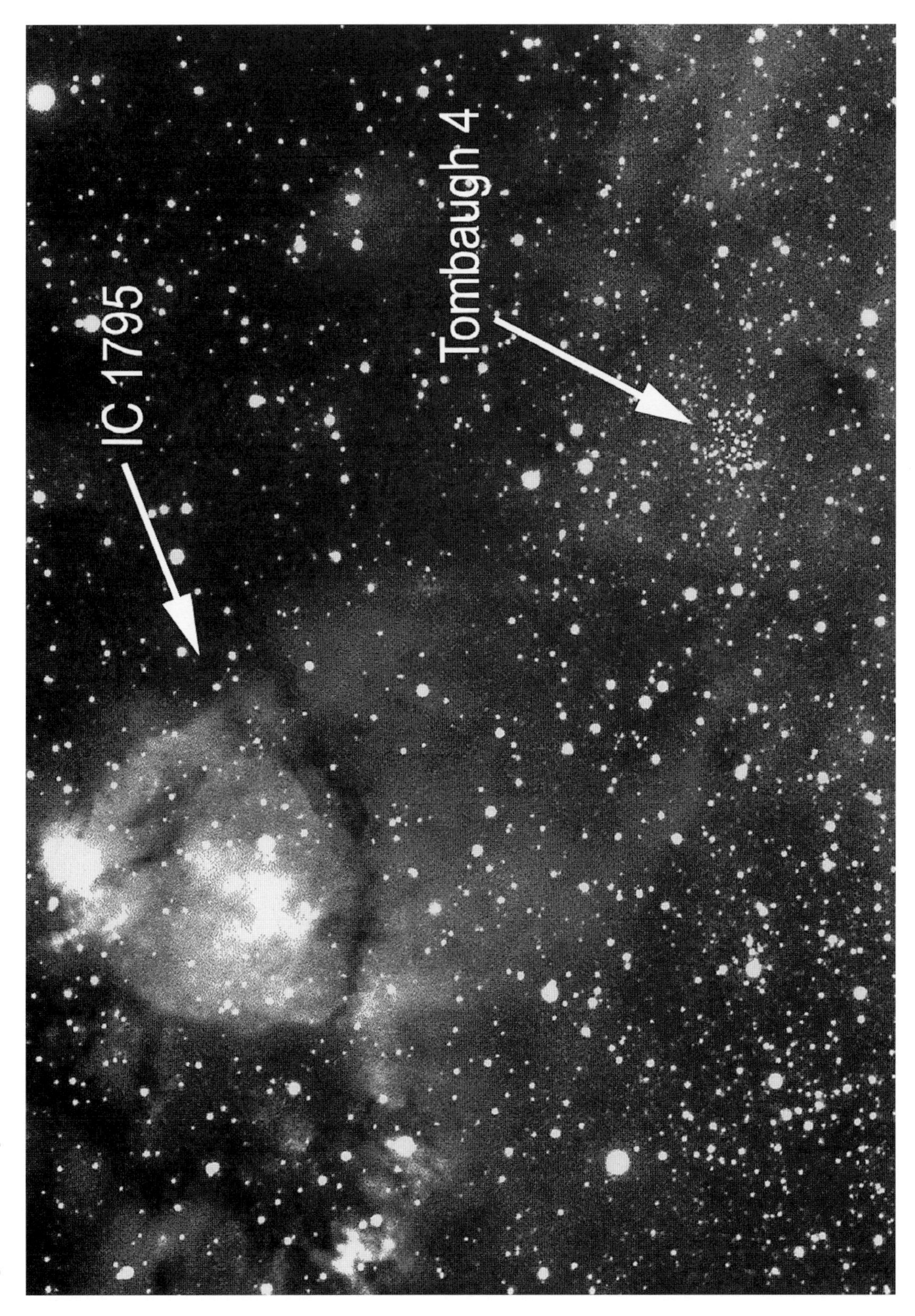

Figure 28 Tombaugh-4.

Figure 29 M7.

Figure 30 Berkeley-17 (the oldest known open cluster).

Figure 31 Double Cluster (NGC 869 and NGC 884), in Perseus.

Figure 32 M45 in Taurus (the Pleiades).

Figure 33 NGC 253 (galaxy). and NGC 288 (globular cluster).

Figure 34 M4 and Rho Ophiuchus complex.

Figure 35 M13 in Hercules.

Figure 36 Omega Centauri.

Planetary Nebulae and Supernova Remnants

These two structures in space represent late or final stages in the evolution of stars. About 95% of stars will end their evolution quietly, while the remaining 5% will detonate in violent supernova outbursts. The critical feature which determines a star's final path is its mass. Stars unable to get below eight solar masses by the time their nuclear fusion reactions end will go supernova, while one member of some binary stars will explode as an end-point of a process involving mass exchange. These types of supernova are known as type II and type I respectively.

The photograph of **M51** shows supernova SN 1994i close to its nucleus. This location makes life difficult for supernova hunters, as the nucleus can be easily overexposed and thus hide the suspect object. Here a Celestron 14-inch SCT was used at f/11 with Fuji HG 400 film, and the 90-minute exposure shows the supernova to good effect under the dark skies of Mt Pinos, California.

Supernova remnants (SNRs) such as M1 and the Veil Nebula tend to be more irregular than planetary nebulae and this reflects the violence of their birth. Some of the more massive giant stars may not be able to shed enough material during the later stages of their evolution to get below the size limit for white dwarfs. Above the limiting mass, collapse beyond a white dwarf will take place, producing a supernova explosion and an SNR around it. Depending on the nature of the progenitor, the SNR may be filled (called a plerion, with a neutron star – pulsar – or a black hole left behind) or not. At present the majority view is that all single, massive stars will evolve to undergo supernova explosions, producing remnants. Type I supernovae, involving a star in a binary system, leave no collapsed stellar remnants.

Black holes are one of the most enigmatic structures in space, because by definition they can never be seen, their presence can only be inferred. If an object which is not visible at ordinary wavelengths is nevertheless a strong X-ray source, and if the mass can be estimated through observations of a visible companion object, the presence of a black hole can be inferred if the mass of the invisible object is above the theoretical limit for a neutron star, currently thought to be around twice the mass of the Sun.

The X-rays mentioned above derive from the gravitational energy of matter in an *accretion disc* around the black hole, orbiting, compressing, heating up, and falling into oblivion. At the centre of a black hole, there is a *singularity*, a point of zero size and infinite density where matter may be crushed out of existence. One well-known candidate for a black hole is **Cygnus X-1**, the most powerful X-ray source in the constellation Cygnus the swan. When looking at Michael Stecker's image, remember that it is the visible companion to the supposed black hole which has been captured in print.

The neutron star phenomenon is barely less exotic. Given that our Sun is approximately 800,000 miles in diameter, the white dwarf it leaves behind will be about the size of the Earth at 8,000 miles in diameter. A neutron star represents a state of matter not unlike a giant atomic nucleus, where the fundamental constituent particles of atoms are compressed very tightly indeed, so much so that protons and electrons cannot survive. Only neutrons – and perhaps some even more exotic "heavy" particles –

can be found. Such an object will have the mass of an average star crushed to a diameter of only 80 miles, or much less. The surface of neutron stars must be very smooth, due to the immense self-gravity of such objects.

Most of the light emitted from neutron stars is released by some sort of beaming mechanism, giving a lighthouse effect as the central star itself will be rotating at a rapid rate. Like the spinning skater pulling in his or her arms, a shrinking star will spin up faster and faster. The fastest rotating neutron stars have rotation periods of the order of milliseconds, and their rapidly flashing pulses of radiation need very short timescale observations to be resolved. This effect, observable at radio as well as visible wavelengths, gave rise to the name *pulsar.*

The Crab Nebula (M1 in Taurus) is possibly the most famous SNR and probably the most frequently observed object in the skies. We are fortunate in our galactic geography since the focused beams of radiation from the Crab neutron star sweep across space towards the Earth and so allow us to study the beaming mechanism. The event which produced the Crab was recorded in July of the year AD 1054 by Chinese astronomers as a new, bright "guest" star near one horn of the constellation Taurus the bull. Several centuries later John Bevis' telescope – trained on the same patch of sky – showed up the Crab Nebula for the first time. Shortly afterwards Charles Messier observed this object, while locating the comet of 1758 which travelled between the bull's horns. The Crab Nebula was given first place (M1) in Messier's catalogue.

Other supernova remnants visible to the camera and CCD chip include **IC 443** in Gemini, responsible for a serendipitous comet discovery some years ago, together with **S147**. S (Simeis) 147, also known as Sh2-240 is a huge and very faint SNR about 3 degrees in diameter, equal to six full moons. It is thought to be just over 1 kpc from Earth. S147 is to be found 5 degrees north-north-east of the much brighter and smaller SNR, M1 the Crab Nebula in Taurus. Like M1, it is associated with radio (synchrotron) and X-ray radiation. The radio source for S147 is the 143 millisecond pulsar, PSRJ0538+2817. This pulsar is located 40 minutes west of the centre of S147. Unlike, M1, S147 is much larger and therefore – given a similar distance – older with an estimated age of 80,000–200,000 years.

The optical appearance of S147 is that of a large ball of thread-like filaments. It is rarely seen or photographed because of its very low surface brightness. This SNR was named in part for the location of the observatory where it was discovered in 1952: the Crimean Astrophysical Observatory at Simeis, USSR. Stecker's image of the south-west half of the SNR was the result of a two-hour exposure using a 130 mm reflector at f/4.5 through a red (W25A) filter on hypersensitized pre-treated Kodak TP 2415 film.

A much younger SNR, IC 443 in Gemini, has an estimated age of 1,000 years. It is smaller than S147 but larger than M1. IC 443 is situated close to the bright star ETA Geminorum. Although it has a similar filamentary structure to S147, it is relatively asymmetric. This asymmetry is thought to be caused by its interaction with a molecular cloud, or possibly a second supernova explosion in the past. The black and white photo of IC 443 was taken with a 10-inch f/4.5 telescope. The exposure was for 100 minutes through a red (25A) filter on hypersensitized Kodak TP 2415 film.

Supernova remnants can contain ten solar masses of material or more, and expand at high velocities, up to 30,000 km/s initially. Photographs of the filamentary structure of the Crab Nebula taken some tens of years apart reveal the continuation of this expansion. Some large, very old SNRs have been found, since the visible lifetime of the debris is estimated at around 100,000 years. Much depends on the location of the event and what lies between it and the Solar System, and some SNRs have only ever been observed at radio or other non-visible wavelengths.

The largest SNR is not visible from northern hemisphere skies. Named the Gum Nebula, after the Australian astronomer Colin Gum who discovered it, this was initially thought to be hydrogen made visible due to ionization by two bright stars nearby. Later, a theory was put forward that the ionization had been caused by a burst of radiation from a supernova many thousands of years ago. In agreement with this idea, there is a pulsar close to the centre of the Gum Nebula.

Another elderly remnant is the **Veil Nebula** in Cygnus (NGC 6960/6992/6995), thought to be the relic of an event which occurred about 100,000 years ago. As with the symmetrical planetaries such as the Ring Nebula, we see arcs or circles rather than shells in many cases, since we are looking through a greater thickness of material at the edge of a bubble than in the middle. The Veil can be glimpsed visually – once managed by this writer unintentionally, while setting up a photograph using a low-power 80 mm aperture finder-telescope – on a very dark night. High-quality visual filters can help redeem the situation if the observation is being made under light-polluted skies.

Supernova explosions produce intense bursts of radiation, including exotic particles such as neutrinos together with more mundane electrons, protons and neutrons. Neutron capture during supernova explosions is an important process in terms of the production of the chemical elements heavier than iron and nickel. Nuclear fusion inside stars continues to the point when there is no longer any energy to be had from joining smaller atomic nuclei together to make bigger ones, which happens to coincide with the production of elements such as iron. Beyond, this neutron capture in a supernova followed by radioactive decay, will generate elements such as silver and gold. The jewellery which many of us wear on our fingers contains atoms with a very violent history.

Lower-mass stars like the Sun will expand into red giants before ejecting their outer layers of gas into space in a more controlled and less energetic process which produces what is known as a planetary nebula. Planetary nebulae are infamously misnamed structures in space. Appearing as small discs, some of the more symmetrical examples do resemble a planet as viewed at low magnification, but this is where the connection ends. Planetaries are isolated gaseous nebulae consisting of a shell of gas and dust that has been ejected in the recent past (astronomically speaking) from an evolved central star. This star is, or soon will be, a white dwarf, and is hot with a surface temperature of 50,000° to 300,000° Celsius. In effect, the central white dwarf star in a planetary is the cooling exposed core of its former giant self.

Measurements made using the wavelengths of light emitted from planetary nebulae show that the shells are expanding outwards at velocities of around 25 km/s, giving a visible lifetime of around 20,000 years. The energy output near the central star is sufficient to ionize many of the elemental gases contained in the shell, and many planetaries contain significant amounts of ionized atomic oxygen, nitrogen and hydrogen. Containing about a solar mass or so of material, the planetary shell is typically a light year or so across. At the centre of a planetary nebula is a star which is compressed to such a degree that a few cubic centimetres of its matter would have a mass of several tonnes. However, this is lightweight stuff compared to the densities achieved by neutron stars, where the same small volume of material could correspond to a mass of billions of tonnes (10^{12} kg).

The only chemical elements produced in large quantities during the Big Bang were hydrogen and helium. Heavier elements synthesized by nuclear fusion inside stars are distributed through the inter-stellar medium partly through the dissipation of planetary shells. These elements contain the essential atomic building blocks of life, such as carbon, nitrogen and oxygen. Second and later generations of stars, and any associated planetary systems such as ours, will contract from nebulae enriched in these elements. In spite of what our teachers may have said, we really are star material.

The nature of the light from planetaries' gaseous shells, involving pure colours of a single wavelength, allows astronomers to make use of high-tech visual and photographic accessories for identification and observation of the smallest and youngest objects. As observational targets they provide challenging and intriguing views. Since the vast majority of the stars we can observe lie in the plane of the Milky Way, this is the place to look for most planetary nebulae and supernova remnants. Any selection of a dozen or so well-known planetaries would show the varying degree of symmetry possessed by such objects, and also the wide range in angular size. Some of the largest and faintest have only recently been identified on wide-field sky survey plates, while some of the smallest are difficult to distinguish from field stars.

Broadly speaking the former category are unrealistic targets under light-polluted skies, though there are ways of improving your chances with both. OIII filters, one of the high-tech accessories mentioned previously, offer the best opportunity to view planetaries under light-polluted skies, as they reject the yellow pollution from common sodium lights but allow through green light from the nebula. Spotting a small planetary against a rich Milky Way starfield can be assisted by using a medium power and a diffraction grating (say 80 to 200 lines per mm) to split light up into a spectrum. When the grating is placed between eye and telescope the light from stars in the field – consisting of a wider spectrum of colours – is spread out and so considerably dimmed, while the relatively monochromatic light from an almost stellar planetary will remain bright in the primary band.

Well-known objects include the small but impressive Ring Nebula (M57) in Lyra, the infamous Dumbbell Nebula (M27) in Vulpecula, M76 (the mini-Dumbbell in Perseus), **M97** (the Owl Nebula in Ursa Major), NGC 6781 (Aquila), **NGC 7293** (the Helix Nebula) and NGC 7009 (the Saturn Nebula in Aquarius).

Many of the more symmetrical objects appear as circles of blue-green light, with red in some planetaries due to ionized hydrogen. These discs are actually three-dimensional shells, in the manner of a soap bubble, and in some cases spherical symmetry is lacking, probably as a result of the rotation of the parent star. However, recent observations of the Ring Nebula using the Hubble Space Telescope suggest that it is cylindrical rather than spherical.

In looking at photographs or electronic images of these planetaries, remember that different emulsions – and CCD chips – give different responses to the human eye. With some of the more symmetrical shell objects, the presence of red (hydrogen) in the outer regions and green (oxygen) closer to the central star reflects the lower ionizing power of radiation as distance from the hot white dwarf surface increases. It is unlikely that the eye will see such shades of colour in most planetaries using moderate apertures, and green alone will most likely appear in larger instruments.

Figure 37 Planetary nebula M97 and galaxy M108.

Figure 38 IC 443 in Gemini.

Figure 39 Simeis-147 (Sharpless-240). – southwest portion.

5 Galaxies and Galaxy Groups

Before stellar evolution theory reached its present stage, galactic evolution ideas were tentative and depended not so much on physical arguments (e.g. galactic dynamics) but intuitive notions based on observations of galaxy shapes. Historically this is quite understandable. The first such theory was due to Hubble, and relates to the Tuning Fork diagram classification of galaxies as irregular, elliptical, spiral or barred spiral. Hubble's theory was that galaxies began as large spherical objects (ellipticals) which then flattened due to rotation, becoming increasingly eccentric before transforming into a tightly wound spiral. Further flattening and loss of material from the nucleus produced the more open spirals, with almost no nucleus and extensive spiral structure. Eventually a final stage is reached where all structure disappears and an irregular galaxy remains.

Interestingly, another astronomer, Shapley, proposed that exactly the reverse sequence was operating. Within our present understanding of how stars evolve, both of these ideas fall. Being able to estimate, with some degree of confidence, the ages of stars has revealed that the galaxies – of all types – which hold them are of approximately the same age (of the order of ten thousand million years). While the oldest stars in galaxies are of the same age, we find that elliptical galaxies contain largely these old stars, spirals have a mixed population of young and old, while irregulars hold mostly young stars.

What the Hubble classification may be showing us is not so much an evolutionary sequence as a conservationary one – how good certain galaxy types are at conserving their star-forming material. Elliptical galaxies seem to have exhausted their supply with a burst of early, rapid star formation, while the irregulars have been much more conservative. Present theories of galaxy formation suggest that the initial collapse of, and subsequent star formation in, a proto-galaxy are particularly strongly influenced by the original rotational velocity of the proto-galaxy, amongst other factors.

Proto-galaxies with moderate to little angular momentum, allow rapid early star formation and produce ellipticals, while rapid rotation appears to generate galaxies with a more steady rate of starbirth. The distribution of globular clusters in galactic haloes could be a consequence of the more spherical shape adopted by spiral galaxies like our own, in the early stages of collapse and star formation. Subsequent evolution is probably influenced by violent events within galaxies as well as by internal and external gravitational effects.

A puzzle involving the structure of ordinary spiral galaxies concerns the existence of spiral arms. Galaxies are rotating – our own Milky Way takes 225,000,000 years to rotate once and complete a "cosmic year" – and after several rotations the spiral arms should wind themselves up and disappear. Yet there are many spiral galaxies older than a few rotations which maintain a very distinct and beautiful spiral shape. A rotating density wave could be responsible. In this model, material in a galactic disc periodically enters and then leaves the region of higher density. On entry, the gas and dust is compressed, triggering star formation.

This fits well with observation, as spiral arms are where nebulae and young stars are situated. Some galaxies, such as M63, may have spiral structure caused by supernova explosions. The bar extending from the core of "barred spiral" galaxies is another structure in space which remains to be fully explained. Examples of this type of galaxy include M109, an object in Ursa Major, and NGC 7479 in Pegasus. Other deep sky astronomers' favourites include **M33** (the Pinwheel Galaxy in Triangulum) and M74 (Pisces), two spiral galaxies which present a face-on aspect and so reveal their spiral structure very clearly, with nebulae and clusters also visible to those using large telescopes. The M33 image resulted from a 45-minute exposure through a 200 mm f/4 reflector on hypered Kodak TP 2415 film, and shows a wealth of structural detail in this near neighbour of the Milky Way.

M51, the Whirlpool Galaxy in Canes Venatici, shows distortions in the spiral arms due to the gravitational interaction with its neighbour, a smaller but more massive elliptical. A seemingly bright field star near the nucleus of this galaxy is in fact a supernova, something which both professional and amateur astronomers invest much time searching for, and which has been discussed in detail earlier in Chapter 4. A number of galaxies have spiral arms which, instead of emanating from the nucleus, begin from the ends of a bar which extends from the nucleus. When observing these barred spiral galaxies, it is sobering to consider that astronomers have not yet explained satisfactorily the mechanism behind the origin and maintenance of the bar structure.

Some of the most impressive sights are those provided by edge-on spirals. Here the cold, dark matter in the spiral arms appears as a thin dark line bisecting a needle of light. Popular examples are **NGC 4565** in Coma Berenices, **NGC 891** (in Andromeda) and **M104** (the Sombrero Galaxy in Virgo), galaxies imaged by Michael Stecker and featured in the CD-ROM selection. The dark equatorial band seen in these galaxies represents the dark molecular clouds of star-forming material exactly like those we see in our own Milky Way Galaxy, and indeed as seen in Michael Stecker's images in Chapter 2.

For those seeking the ultimate challenge, try finding the 18th magnitude galaxy UGC 2838 (see page 47) near M45, an object mentioned previously and clearly visible edge-on in Stecker's CD image of the Pleiades. In addition to observing detail in galactic structure, ranging from visibility of arms in a spiral structure to individual bright nebulae or star associations and clusters, systematic galaxy observation may produce a supernova discovery.

More challenging still, and potentially very fruitful as a line of research for both professional and amateur astronomers, are the enigmatic active galactic nuclei (AGN), a category of objects which includes quasars (QSOs) and BL Lacertae objects. As the name suggests, AGNs are thought to be galaxies with particularly bright and energetic nuclei. Quasars were once thought to be stars, as they are compact objects with no resolvable detail in ordinary images – they look just like stars. Observations made by splitting the light from quasars into a spectrum have shown that they are likely to be at very great distances, and therefore cannot be single stars as these would be invisible at the distances involved.

These objects are now believed to be the cores of very distant galaxies, where a young black hole is siphoning stars and releasing huge amounts of gravitational potential energy. Many are strong sources of radio waves, and lists of AGN frequently include designations such as 3C66A, where the "3C" refers to the *Third Cambridge Catalogue of Radio Sources*. Other designations in common use include PKS (Parkes Radio Observatory). These objects make the distance to M31, the Andromeda Galaxy, seem like a short hop. The light emanating from quasars started its journey when the Universe was a fraction of its present age.

Other types of AGN have nuclei which are less energetic but equally puzzling, such as Seyfert Galaxies as illustrated by M77 in the constellation Cetus the whale. Some are the subject of continuous interest, particularly 3C273 in Virgo, while others are less well known but of similar importance such as Mk (Markarian) 205, situated near the much closer NGC 4319 but apparently linked to it, Mk 421 (near the bright field star 51 Ursae Majoris), and OJ 287 a BL Lac object in Cancer.

The disturbed looking **M82** in Ursa Major has been the subject of much debate, astronomers disagreeing about the origin of its intense activity. There appears to be a high energy outburst producing two lobes of electromagnetic emission at 90 degrees to the arms of this edge-on galaxy, and the nucleus itself looks irregular and troubled, appearing mottled through moderate aperture telescopes with a dark absorption band running at an angle across it.

Given the tendency of galaxies to occur in clusters, interactions between the gravitational fields of galaxies will be fairly common, and it appears that there have been mergers between galaxies as a result of direct collisions. Today, some nearby giant elliptical galaxies show double or even triple concentrations of stars near their centre, a feature probably due to the merger of two or more spiral galaxies, giving rise to multiple galactic nuclei. These encounters appear to strip the participating galaxies of much of their gas and dust.

Where the collision is more oblique, the changes wrought in the shape of the galaxies are varied and fascinating. Some encounters generate "sprung spirals" where the spiral arms lose their circular form and extend further away from the core, perfectly illustrated by the Whirlpool Galaxy M51 which was discussed earlier in this chapter. Other angles of approach produce rat-tail galaxies, where stars are strung out in long tails.

A head-on collision between one small galaxy and a much larger one can give rise to a cartwheel shape in the larger of the two galaxies. The direct passage of an interloper galaxy through another may not result in a single stellar collision; because of the large amount of empty space in each, head-on crashes between stars are unlikely, but the force of gravity does re-shape both protagonists. The larger galaxy ends up with its core surrounded by a bright outer ring, probably a manifestation of a pressure wave where star formation is triggered, together with some spoke-like structures radiating out from the core. The smaller galaxy is likely to be stripped of star-forming gas and dust, and remain barren thereafter.

If a telescopic image of some parts of the sky is examined in detail, large numbers of galaxies can be seen, not as magnificent spirals as depicted on photographs, but as small fuzzy patches of light. One constellation with more than its fair share of galaxies is Virgo, but there are several others including Ursa Major, otherwise known as the Plough or Big Dipper. Observations across the sky confirm that the distribution of galaxies is far from uniform on this scale, and that galaxy groups or clusters are quite common. Indeed our Milky Way Galaxy is one of two prominent members of a group of galaxies called, rather predictably, the Local Group (see Table 3).

Table 3. The members of the Local Group. From the column showing distance it can be seen that four of those galaxies listed are satellites of the Great Andromeda Spiral Galaxy, M31

Galaxy	Type	Distance (kpc)	Mass (Sun = 1)
M31	Spiral	710	3×10^{11}
Milky Way	Spiral	–	2×10^{11}
M33	Spiral	730	1×10^{10}
LMC*	Irregular	50	1×10^{10}
NGC 205/M110	Elliptical	710	1×10^{10}
IC 10	Irregular	1250	4×10^{9}
M32	Elliptical	710	3×10^{9}
SMC*	Irregular	60	1×10^{9}
NGC 185	Dwarf elliptical	710	1×10^{9}
NGC 147	Dwarf elliptical	710	1×10^{9}
NGC 6822	Irregular	470	4×10^{8}
NGC 1613	Irregular	740	3×10^{8}

*LMC/SMC = Large/Small Magellanic Cloud

These agglomerations are very large indeed, with diameters of 10^{20} km or more, and with hundreds of members. A typical example might have a mass of 10^{45} kg (equivalent to some 10^{15} Suns) and 200 member galaxies. The average mass of a galaxy – 5×10^{12} solar masses – is then ten times the mass of the Milky Way. There are many spirals bigger than our own, but some of the extra mass will be invisible cold, material between the galaxies, which will not show up on photographs.

Other galaxy clusters can be seen out to great distances. Some, such as **Stephan's Quintet** (located near – in a south-westerly direction – the more prominent galaxy **NGC 7331** in Pegasus), are very popular targets for amateurs, while others listed in the *Abell Catalogue of Galaxy Clusters* are difficult targets without a large observatory telescope to use. Stecker's image of NGC 7331 and the Quintet shows the large galaxy to be something of a distant cousin to M31, the Andromeda Galaxy. Both have the same general form, with satellite galaxies in attendance.

Stephan's Quintet consists of galaxies designated NGC 7317, 7318A and B, 7319 and 7320. There are many puzzles surrounding this famous galaxy group. On deep observatory photographs there appears to be a filament of luminous gas enveloping the Quintet which extends across to NGC 7331 – although the latter appears to be much closer, making a physical bridge such as this unlikely. In addition, one of the group (NGC 7320, the largest member) appears to be receding from the Local Group at a far slower rate than the other members. A possible explanation is that Stephan's Quintet is moving apart, a scenario in which it is reasonable that at least one of the galaxies should be moving towards us in relative terms, with this motion partly counteracting the recession of the group as a whole due to the expansion of the Universe.

Although amateur observation of remote galaxy clusters is difficult, it is not impossible. One worth trying for if you have a telescope of suitable size is A2151 in Hercules, and two more potential targets include the Pavo Cluster, visible to observers in the southern hemisphere, and A1060. This pair provides an example of a regular (Pavo) and an irregular galaxy cluster. The deep sky contains a large number of galaxy groups, stretching from the northern galaxies of Ursa Major and Canes Venatici, Hercules, and the rich Leo/Virgo region which is accessible from both hemispheres.

There are several types of star which vary their output in a regular or semi-regular way. Many such stars can be found in the Milky Way, and some of the brightest can also be seen from great distances across space. This allows them to be used as "standard candles" to estimate the distance to their parent galaxy. Such stars have been identified in M31, which is over two million light years from the Milky Way. Even at this great distance, spotting the globular clusters of M31 is possible with modern amateur telescopes and provides a challenge which attracts many followers.

The Hubble Space Telescope is currently providing some excellent images of galaxies at the very edge of the observable Universe, providing a view back in time near the epoch when galaxies were formed out of the material of the Big Bang. This represents an era where spirals seem to be more predominant than nowadays, with clusters of galaxies otherwise much like our own era. The presence of clusters of galaxies is not the end of the line in terms of structures in space, however. The overwhelmingly large number of galaxies seen in the direction of Virgo is a sign that here is a supercluster of galaxies, one of the largest structures in the Universe. Stecker's image of the **Virgo Cluster** is centred near **M86** and shows many objects along and beside the "Markarian Chain". It was taken using his 155 mm f/7 refractor and a 60-minute exposure on hypered Kodak TP 2415 film. The resulting image was scanned and further processed using Adobe Photoshop software to enhance detail and contrast. M86 and M84, the two prominent elliptical galaxies in Stecker's image, are visible with small amateur telescopes.

One of the key questions facing mankind, let alone astronomy, is the issue of whether the Universe will continue to expand forever, or whether the mutual gravitation of its contents is sufficient to halt the expansion and bring the galaxies rushing back towards each other leading to what has been described as the Big Crunch. At present it appears as though the amount of matter in the Universe is very close to the critical value, and the jury is still out.

There are theoretical reasons for expecting the value to be exactly on the borderline between these two outcomes, and the puzzle is set to keep astronomers busy for some time yet. In some ways it would be a shame if observational astronomy was in some way able to estimate the total mass of the Universe sufficiently accurately to answer this key question. Somewhat perversely perhaps, I hope that the answer will remain elusive, although the issue will remain literally academic for many generations.

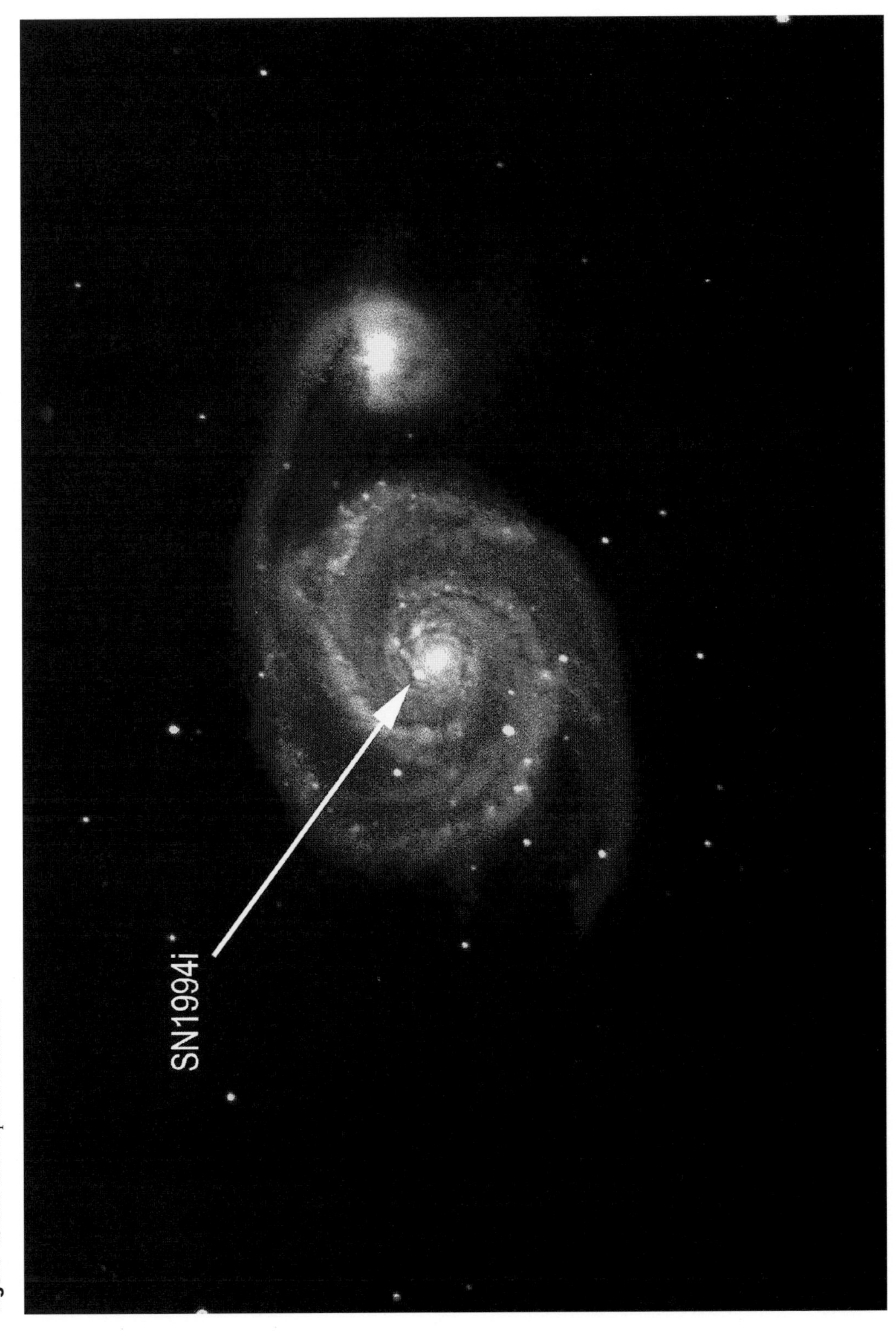

Figure 40 M51 with Supernova SN1994i.

Figure 41 Virgo Cluster (Markarian's Chain).

Figure 42 M33.

Figure 43 NGC 891 in Andromeda.

Figure 44 M104, the Sombrero Galaxy in Virgo.

Figure 45 Virgo Cluster of galaxies.

Figure 46 M65 and M66 in Leo.

6 Observing Notes

Observing or imaging deep sky objects is not particularly difficult at the most basic level, but like any skill it improves with experience and is always helped by high-tech equipment when used appropriately. It is also helpful to have a good knowledge of the sky, as most of the nebulae listed in the various catalogues are below naked eye visibility. The best approach to gaining first-hand knowledge is to join a local astronomical society. The brightest nebulae, such as M42 and M8, can be seen with the unaided eye on a clear night away from city lights. The rest require binoculars or a telescope.

A rich collection of easily located starfields plus dark clouds (and bright emission nebulae) can be found in the constellations of Scutum and Sagittarius. To try and capture star fields plus dark nebulae on film, try 20–30-second stationary camera exposures using an SLR camera fitted with a 50 mm or smaller focal length (wide-angle) lens and fast (ASA/ISO 400 to 3200) film. Use a sturdy tripod and a cable to release the shutter (set to "B" for a time exposure).

Colour transparencies are least susceptible to commercial lab processing errors if, often as a beginner to this type of photography, you cannot develop and print your own work, although it is a good idea to start each roll of film with a standard terrestrial photo to allow technicians to cut the film accurately after development. Without this step it has been known for films to be returned with each frame neatly snipped down the middle. Taking colour print film to the photoshop can cause problems, since machine operators rarely know how to set up the equipment for "weird" astrophotos, and the resulting prints often emerge looking like a close-up of a muddy puddle.

For good, dark skies limit your exposure – in minutes – to something between twice and four times the square of the focal ratio you use. At the top end of this scale, the longest exposures take into account the use of filters. So at f/4 the maximum exposure using the typical film speeds quoted would be 30–60 minutes depending on the actual sky conditions at your site and whether a filter was used. In heavily light-polluted skies, between one half and one quarter of this time is safer, maybe less. This rule of thumb can also be extended to other types of deep sky photography. Experimentation is advisable, increasing the exposures in increments with all times recorded for evaluation when the results can be inspected. The aim is to reach the limiting exposure where skyglow (or light pollution) just fails to degrade the image.

It isn't immediately obvious how much of the sky will appear on your film. Table 4 shows the angular field of

Table 4. Fields of view for lenses of various focal lengths. The fields available through a telescope can be estimated by extrapolating from these figures, as a doubling of the focal length halves the field of view

Lens focal length (mm)	35 mm frame field (degrees)
50	28 × 42
135	10 × 15
200	7 × 10.5
300	4.5 × 7
400	3.5 × 5
500	2.8 × 4.2

view obtained on a 35 mm film frame using various telephotos. For comparison, the full moon is approximately one half of a degree across, so a field of view 10 × 15 degrees will be 20 by 30 moonwidths.

To assist with sketches of any extended deep sky object, try mentally dividing the field of view into quadrants, as this will help with positional detail. Averted vision will bring fainter parts of a nebula within reach. Sketching on to tracing paper taped over a red torch face (if flat enough) will prevent loss of dark adaptation, and if the tracing paper has a circle divided into quadrants then transcription may be further eased. Some observers with DIY skills have taken this idea one stage further by inserting a transparent plastic disc, etched with a grid system, into the focal plane of an eyepiece.

This arrangement will provide the basics of a coordinate frame of reference when illuminated from the side by a red LED in the manner of a guiding eyepiece. The nature of the light from emission nebulae allows a high-tech solution to the problem of light pollution. Since the spectral lines characteristic of emission nebulae are mostly situated away from light pollution lines (sodium, and some mercury), interference filters can be used to remove the light pollution selectively. Some attenuation of nebular light does occur, but the gain in signal to noise ratio can bring previously invisible objects into view.

For visual observers the O III filters can produce excellent results, particularly with planetary nebulae, while photographers will find the less specific deep sky filters beneficial. Choosing an eyepiece for deep sky observing can be as much a matter of cost as suitability. I have obtained excellent views with relatively inexpensive Plossls and Erfles. The wider field of view obtained with Naglers can be spectacular but more light loss is bound to occur with the larger multi-element designs.

In scanning literature and catalogue references to nebulae, magnitude values quoted can be confusing and misleading as an indicator to visibility. Integrated magnitudes do not give as true an indication of appearance as visual surface brightness magnitudes, which are rarely given – the latter take into account the area over which the brightness is spread. This is another thread of the reasoning behind our warning, given much earlier in this book, about visual views through a telescope sometimes being disappointing.

To gain an impression of the visibility (or otherwise) of an object prior to observation, it may be better to ask somebody who has seen it than to consult magnitude lists. Photographically, emission nebulae provide some stunning views of the heavens, as can be seen from the images in this book. Exposures of only a few minutes' duration can reveal impressive amounts of nebulosity. As a compromise between light grasp, cost and manageability, the traditional 10 or 12 inch f/5 or f/6 reflector has much to recommend it, while modern computer-controlled Schmidt-Cassegrains are a wonderful tool for the informed (and especially the IT literate) amateur.

With a well-aligned equatorial, skilful guiding and an appropriate choice of film and processing, views can be obtained which will rival professional photographs. Black and white films such as Tri-X and T-Max were firm favourites for many years, particularly under the light-polluted UK skies, when combined with an energetic developer such as D-19 or MWP-2. Under better conditions, long exposures using hypersensitized Kodak TP 2415 modern high-speed film will always lead to better results due to its greater sensitivity to red light and superb grain characteristics. Colour emulsions have much to offer, particularly when hypersensitized.

For optimum results, photographers use a variety of techniques to improve the performance of their films. Cold camera techniques have been used to good effect, as demonstrated by the well-known Canadian astrophotographer Jack Newton, although these days the camera undergoing cooling is usually of the CCD variety. Techniques such as latensification, treating the exposed emulsion with reagents including mercury vapour, can also be used to improve the final product. Darkroom techniques available to the amateur such as "dodging" and "burning" images, are outside the scope of this book.

One approach, which mimics the digital technique of stacking images using photo-processing software, is to carry out the physical equivalent with two negatives in an enlarger. By carefully aligning and "stacking" two negatives of the same object (taken consecutively on the same night to improve ease of alignment) in the enlarger, contrast can be improved by over 40% since the grains in the film will cancel out to some degree whilst the image will be reinforced. The much longer exposure times needed to generate a decent print – together with the difficulty of accurately stacking three or more negatives – are usually sufficient deterrents to extending this technique beyond two negatives.

The advent of computer-controlled telescopes and electronic imaging systems (CCD cameras and software to process the raw images) has replaced photography to a large degree, although wide field views are still the province of camera and film as CCDs – at least those available within amateur budgets – remain fairly small and incapable of covering much more than a fraction of a degree of sky when used with a telescope.

However, with larger telescopes the field of view is restricted anyway, as can be seen from an extrapolation of the values in Table 4 to the focal length of, say, Stecker's 14-inch (356 mm) f/11 SCT. With a 500 mm telephoto lens, the field of view is 4.2 × 2.8 degrees on a 35 mm format film. At a focal length in excess of 3,900 mm, the field drops to around 0.53 × 0.36 degrees, so that a full moon no longer fits in the frame. With some CCDs, the field of view will be something like 10 minutes of arc in diameter (60 minutes of arc = 1 degree).

The use of CCDs will nevertheless bring a world of faint structures in space within the grasp of the amateur astronomer. The rules of the game change dramatically when using a CCD. Operating at low temperatures, commonly –20° Celsius, they detect low intensity light much more efficiently than a photographic emulsion. While a camera plus film will record only a few percent of the photons of light hitting the film, a CCD will be 10 or 20 times more efficient. As a result, short exposures reveal impressively faint objects, while long exposures reach objects previously only targeted by professional astronomers with access to giant telescopes on mountaintop observatories, as long, that is, as the CCD doesn't become saturated (overexposed) and the amateur's telescope has been aligned and has tracked the target across the sky accurately enough in the first place. Surprisingly perhaps, Michael Stecker's images were captured without the use of a CCD – here's how.

To enhance analogue (film) images digitally requires that the negatives first be scanned, or digitized. Scanning techniques are variable and the quality of the final image is dependent on the quality of the scan. A digitized Stecker image can then be enhanced by computer (nowadays, a Pentium-II 300 MHz with 12 GB hard drive, 256 MB of RAM and Windows 98 operating system). The two-image processing software he uses are Picture Window 2.0 or 2.5 (http://www.dl-c.com) for stacking and Adobe Photoshop 4.0 (http://www.adobe.com) for most of the other image processing.

One of the most powerful methods he uses for enhancing images is stacking or compositing of multiple negatives. This process increases the signal to noise ratio, equivalent to reducing the graininess, by a maximum of the square root of the number of negatives stacked. For two images, the square root of 2 is 1.41, and this shows that a 41% increase in contrast is possible. Three images would give a ratio of about 1.7, or 70% increase. Therefore, in order to increase the signal to noise ratio Stecker takes separate photographs of the same object at the observing site and stacks them in either the darkroom as outlined above or, more frequently, with the computer. Digital images can easily be aligned and stacked (called blending) with the aid of Picture Window software. Once stacked there are many ways in which he can further enhance the image with Adobe Photoshop software. It is not within the scope of this book to discuss them further here, but the various techniques include histogram analysis and adjustment; brightness and contrast adjustments, which can be used in a similar way to the burning and dodging methods from darkroom, only much more precisely; colour saturation; blurs; sharpening; unsharp masks; and cloning (to remove artefacts, such as aircraft trails). Readers wishing to explore the world of CCD astronomy in more detail could do no better than read *The Art and Science of CCD Astronomy* and *Software and Data for Practical Astronomers* both by David Ratledge (and also published by Springer Verlag).

So, to sum up, small objects plus visually impressive wide-angle views (the latter impossible to obtain at present using CCDs as their fields are too small) can be produced by scanning analogue images obtained using telescope or lens and a photographic emulsion, or sometimes three emulsions in the case of tri-colour techniques. Image processing software and powerful home computers such as Adobe Photoshop, a Mac Quadra or a Pentium PC are then used to enhance the image which is stored on disc or converted back to either a transparency or negative, from which high-quality prints can be made. That the results are worthwhile can be seen in the Stecker Files, that is, the images used in this book.

7 The Stecker Files

The first time I was contacted by Dr Michael Stecker proved to be quite an eye-opener for somebody in my position, struggling to explore the deep sky through heavily light-polluted UK skies. At that time I was Director of the Deep Sky Section of the British Astronomical Association, author of a regular column on deep sky observing in one of the UK's astronomy periodicals, actively observing (mostly involving a distinctly unfruitful supernova patrol) and occasionally writing about my love of observational astronomy.

Michael had read an article of mine and was interested enough by it to make contact in August 1991. A selection of astronomical photographs accompanied his letter, and I was astounded by their quality. Information contained in an astrophotography article I had seen earlier in *Deep Sky* magazine mentioned the clear skies to be found on Mt Pinos, California, and I replied by asking if this was the site he used. Michael's reply reflected a degree of surprise that I should be able to deduce the site correctly just by looking at the resulting images, but it wasn't that difficult. In some of the communications between us Michael commented on the effects of a volcanic eruption on sky transparency, yet the images accompanying his letters remained of the highest quality, a testimony to the technique as well as the site.

Michael uses kit which includes a Celestron 14-inch SCT, Takahashi Epsilon 200 f/4 reflector, Astrophysics 155 mm f/7 and 130 f/6 refractors, Genesis 102 mm f/5 refractor plus a variety of Tamron lenses (300 mm f/2.8, 180 mm f/2.5 and 90 mm f/2.5 Macro). For tracking he uses an Astrophysics-900 equatorial mount and a Santa Barbara Instrument Group ST-4 autoguider. His prime observing site is the 8,300-foot Mt Pinos, a two-hour drive north of Los Angeles, one of the most popular sites for deep sky observing in the USA.

Many readers will doubtless have seen several examples from the picture selection included here in both my own columns and those of other writers in astronomy magazines around the world. They are of professional quality in many instances and are unequalled in terms of their source for the way they illuminate the structures of the deep sky. To achieve this standard we have seen how Michael frequently uses both analogue and digital techniques in a combination which gives the best of both worlds, allowing him to escape the limitations due to small CCD fields when using telescopes and the added degree of difficulty operating them at remote sites. This interest (arguably a passion) in astronomical imaging wasn't something which hit home at an early age, however.

Born in Brooklyn, New York, in 1943, Michael Stecker grew up in Detroit, Michigan. His education proceeded to a Bachelor of Science degree (1964) in Zoology and a Doctor of Medicine (MD) degree in 1968 from the University of Michigan in Ann Arbor. After moving to California in 1969 he took a residency training in Diagnostic Radiology from the University of Southern California, and is now a Board-certified radiologist practising at an HMO in Culver City, California.

Photography was always a hobby, but he did not venture into the world of astrophotography until around the time of the 1986 apparition of Comet Halley. Much of the time, various pieces of astronomical equipment are used from Mt Pinos (100 miles north of Los Angeles), but Michael has also photographed from Siding Springs Observatory, Australia, Carnegie Las Campanas Observatory, Chile, and Lake Titicaca, Bolivia. Hundreds of his astro-images have been published in books and magazines in England, France, Italy, Japan, Indonesia, Australia and the USA.

A member of the Santa Monica Amateur Astronomy Club and Orange County Astronomers, he has lectured at several astrophotography seminars in southern California and in 1996 spoke on "Enhancement Techniques in Astrophotography" at the European Astrofest in London. Michael now co-sponsors an annual astrophotography award for a UK astronomy periodical, an event which the present writer was once invited to judge (although this never came to pass).

Notwithstanding a busy professional schedule, Michael continues to find time to visit the skies. In terms of targets, he is well known for locating and securing images of some of the more obscure (but fascinating) objects in the heavens. The truth of this is borne out by the fact that, until receiving detailed images from him on Mt Pinos, I and several colleagues had never seen several of Michael's nebulae and clusters (neither at the eyepiece nor on photographic paper or computer screen) until they arrived in the post. Prompted by their beauty and the challenge, many of us tried again – usually with the aid of a deep sky filter and unfeasibly long exposures – and still failed. I am tempted to believe that there is something special about Michael's relationship with the skies that allows him to accomplish feats which mere mortal astronomers in the UK cannot match, in the same manner as the southern hemisphere observer Rev. Robert Evans has a curiously close relationship with supernovae.

In the field of deep sky astronomy, the boundary between art and science is blurred. As an exponent of the art and science of astronomical imaging, Stecker is one of the very brightest stars. He is very successful at what he does and, as we have seen, what he does is quite unusual. Explore the CD-ROM and you'll know what I mean.

Captions to Images on the CD-ROM

0001.jpg

Object name: Solar eclipse – corona
Designation: Sun (colour)
Structures: One of the most impressive sites in nature is a total solar eclipse. The solar corona can be seen at this time. This was the first and only time I have witnessed one. It was in Mexico in the summer of 1991.
Photographic data: Meade 4-inch f/10 SCT. Kodak Ektar-125 film.

0002.jpg

Object name: Meteor and the Milky Way
Designation: Meteor (colour)
Structures: One of the pleasant surprises on a dark clear summer night is a bright meteor display. You do not have to wait for meteor showers. This one was random. It appears to pierce M24 – the Small Sagittarius Star Cloud and the Great Rift of the summer Milky Way.
Photographic data: A Panagor 50 mm macro lens was used at f/5.6 with Fujichrome 1600 film and an exposure of 15 minutes.

0003.jpg

Object name: Asteroid Vesta and the Helix Nebula (NGC 7293)
Designation: Asteroid and planetary nebula (colour)
Structures: On 18 September 1993 Asteroid Vesta passed between our line of sight and the distant Helix planetary nebula in Aquarius. Vesta is the fourth-largest asteroid, at 510 kilometres in diameter. Its surface is covered with basalt, indicating previous volcanic activity.
Photographic data: A 10-inch f/4.5 Newtonian telescope was used with hypersensitized Fujicolor HG 400 film. The single exposure was for 45 minutes on 18 September 1993.

0004.jpg

Object name: Comet Hyakutake (C/1996B2)
Designation: Comet (colour)
Structures: One of the best comets I have seen was Comet Hyakutake. This photo was taken just after maximum brightness on 13 April 1996 from the desert near Red Rock Canyon, California. The comet had a very intense blue ion tail, but little dust. A comet can have both an ion tail and a dust tail. The ion tail is caused by the solar wind and can become distorted and disconnect. The dust tail is composed of larger particles and usually photographs pale yellow instead of blue.
Photographic data: An Astrophysics 130 mm f/6 refractor was used with unhypered Fujicolor 800 film. A single 18-minute exposure was made while guiding on the comet's head. Guiding on the comet is important in rapidly moving comets in order to see maximum detail. Hyakutake did have a large apparent motion because of its closeness to Earth.

0005.jpg

Object name: Comet Hyakutake – solarized
Designation: Comet (colour)

Structures: Comet Hyakutake again, but a few weeks earlier in late March. The comet's ion tail was spread out into several short tails. After taking the picture I scanned the negative so that I could process it in the computer. I used the solarization filter in Adobe Photoshop software to give the comet's head the multi-coloured false-colour appearance you see here.
Photographic data: An 8-inch Takahashi Epsilon 200 reflecting telescope was used with a 35 mm camera and hypersensitized Fujicolor HG 400 film. A single 20-minute exposure was made while guiding on the comet's head. A solarization filter was used in the computer to process the image.

0006.jpg

Object name: Comet Hale-Bopp (C/1995 O1)
Designation: Comet (colour)
Structures: One year after Comet Hyakutake we had another great comet – Comet Hale-Bopp. On 10 April 1997 I photographed it from the desert of southern California. This was the brightest comet I have seen. It had a much stronger dust tail than Hyakutake. Its ion tail was also good, but not as long as Hyakutake's.
Photographic data: A single 20-minute exposure was made with an Astrophysics 130 mm f/6 (telecompressed to f/4.5) refractor and unhypered Kodak Pro 1000 PMZ film. I guided on the comet's head.

0007.jpg

Object name: Comet Tabur (C/1996Q1)
Designation: Comet (colour)
Structures: This is a very interesting comet because it is thought to represent a fragment from a larger comet that had broken apart. Orbital data suggests that at the comet's last perihelion (2900 years ago) it split into two fragments. The larger of the two became Comet Liller (C/1988 A1) which was discovered in 1988 by Dr. William Liller. The smaller fragment became Comet Tabur (C/1996 Q1). The comet's head (coma) has an unusual flattened triangular shape. This comet like Hyakutake was also noted to be emitting x-rays. The x-rays were noted to vary in intensity over short periods of time.
Photographic data: A single 15-minute exposure was made with an Astrophysics 155 mm f/7 refractor and Kodak Pro 400 PPF 35 mm film. By guiding on the comet's head I was able to see its unusual flattened triangular shape and faint ion tail. The photo was taken near Red Rock Canyon, California on October 12, 1996. The image was processed in the computer.

0008.jpg

Object name: The Moon occulting Aldebaron
Designation: Moon (colour)
Structures: On the same night that I photographed Comet Hale-Bopp (10 April 1997) I was also treated to a lunar occultation of the star Aldebaron. In the image you can see Aldebaron (the brightest star in Taurus) a second before it passed behind the Moon. The bright crescent of the Moon illuminated by the Sun is at the bottom of the image. The area above it is partially lit by earthshine, allowing us to see the Moon's darker area as well.
Photographic data: An Astrophysics 130 mm f/6 refractor was used with unhypered Kodak Pro 1000 PMZ film. A single 1/8 second exposure was made.

0009.jpg

Object name: Sagittarius Milky Way and Mt Pinos pine trees
Designation: Star fields (colour)
Structures: The bright red nebula above and to the right of centre is the Lagoon Nebula (M8). Open clusters M6 and M7 can be seen on either side of the middle pine tree. The centre of our Galaxy is thought to lie behind the dark area just to the right of the bright Large Sagittarius Star Cloud seen near the centre of this image.
Photographic data: A single 5-minute exposure (to freeze the trees) was made on unhypered Kodak Pro 1000 PMZ-120 film from Mt Pinos, California. The field of view is 18 × 18 degrees and is centred at right ascension 18 hours 10 minutes and declination –25 degrees. North is up. The image was processed in the computer.

0010.jpg

Object name: Sagittarius Milky Way
Designation: Star fields, nebulae and clusters (colour)
Structures: The dark dust of the galactic equator runs through the centre of the image. Below it (east) and on the right (south) side of the image is the Large Sagittarius Star Cloud. This area seems more yellow-orange and intense than the other star fields and is thought to represent countless numbers of overlapping stars in the general

direction of the galactic centre. There are multiple irregular-shaped dark nebulae insinuating within this star cloud that are difficult to see. An exception is the tiny inky black dark nebula B86 that is well defined and placed just north of the central part of the cloud. In order to see B86 will require that this portion of the image be magnified several times. Just to the west of the Large Sagittarius Star Cloud is the Great Rift and the approximate location of the galactic centre. It is though to lie at right ascension 17 hours 45.8 minutes and declination –28.9 degrees. It also can be located 4 degrees west-north-west from Gamma Sagittarii. The actual nucleus of our galaxies is many light years behind this point and cannot be seen optically because of overlying dust. Just to the left of centre of this image and surrounded by dark clouds is the Small Sagittarius Star Cloud or M24. All along the galactic equator are prominent bright nebulae. The major ones seen here are, from right to left, M8 (Lagoon), NGC 6559, M20 (Trifid), IC 1283/84, M17, M16 and Sharpless-54. To the east of M8 are two globular clusters. The largest and brightest is M22 and the smaller is M28. Less than one degree south-east (lower right) of M8 is the still smaller globular cluster NGC 6544. Planetary nebulae and virtually every other Milky Way object can be found somewhere in this field.
Photographic data: A Pentax 165 mm f/2.8 lens (stopped down to f/5.6) and Pentax 6 × 7 camera was used with unhypered Kodak Pro 400 PPF-120 film. A single 35-minute exposure was made from Mt Pinos, California. The field of view is 18 × 15 degrees and is centred at right ascension 18 hours 10 minutes and declination –20 degrees. North is to the left and slightly up. The image was processed in the computer.

0011.jpg

Object name: Lagoon Nebula (M8) – NGC 6559 – Trifid Nebula (M20)
Designation: Star fields, bright nebula, dark nebula, open cluster and globular cluster (colour)
Structures: Rich star clouds and dark nebulosity outlines two of the most impressive Messier nebulae in the sky (M8 and M20). Besides these nebulae one can also find the open cluster M21 (NGC 6531) at the upper right (north-west) of the image. This is a 5.9 magnitude cluster of several stars that are about 2,000 light years from Earth. At a greater distance is the Trifid Nebula (M20, NGC 6514) which is seen 40 minutes to the south-west. M20 consists of two nebular components. The most northerly is a blue reflection nebula that blends into the brighter southern red emission nebula. Within the southern component are the three dark fissures that give the nebula its name. The probable illuminating star appears as a double in amateur telescopes but is actually composed of six stars. At the bottom centre (south) of the image is the large red emission nebula M8 (Lagoon Nebula). This nebula measures 50 × 40 minutes. Its illuminating star is thought to be 9-Sagittari, seen in its western half. The brightest portion of M8 is a small area called the "hour glass". At the upper left periphery of M8 are two stars. The upper (north) one is GSC6842:1667 (magnitude 8.4) and the lower (south) one is GSC6842:1630 (magnitude 9.8). Between these stars is a "moustache-shaped" area of blue reflection nebulosity that does not seem to be recorded in any of my atlases or catalogues. Extending to the upper left (east) of M8 is very faint red nebulosity that gradually blends into an area of brighter nebulosity measuring about 38 minutes in diameter. This complex consists of a northerly extension fringed in blue nebulosity (IC 1274). Just below it is the red emission nebular complex of IC 1275, IC 4685 and NGC 6559. Traversing this area from north-west to south-east is a river of dark nebulosity called B303. At the east-south-east (left) termination of B303 is a "poppy-like" area of red nebulosity (NGC 6559) with central blue nebulosity around a bright star. The whole nebulous area is well seen because of large unnamed areas of dark nebulosity surrounding the structures. Of interest is globular cluster NGC 6544 which is partially obscured by star clouds at the lower left of the image.
Photographic data: An Astrophysics 155mm f/7 refractor and Pentax 6 × 7 camera was used with unhypered Kodak Pro 400 PPF-120 film. A single 50-minute exposure was made from Mt Pinos, California. The field of view is 2.6 × 2.6 degrees and is centred at right ascension 18 hours 05 minutes and declination –23.8 degrees. North is up and slightly to the right. The image was processed in the computer.

0012.jpg

Object name: Sagittarius and Ophiuchus Milky Way
Designation: Star fields and nebulae (colour)
Structures: The field shows pine trees blurred at the bottom of the image because of the tracking motion of the telescope's mount. At the top and to the left (north-east) is the Small Sagittarius Star Cloud. Below this are the M8 and M20 nebulae. Further below is the Large Sagittarius Star Cloud which lies just east (left) of the point marking our galactic centre. Finally, at the bottom and in line with these structures is a tall pine tree. All these structures lie to the left (east) of the dark central dust lanes of the galactic equator. To the right (west) of the galactic equator are the rich star fields of Ophiuchus with its numerous dark nebulae. The most prominent of these is the Pipe Nebula seen just below and to the right of centre.
Photographic data: A Pentax 165 mm f/2.8 lens (stopped down to f/5.6) on a Pentax 6 × 7 camera was used with unhypered Kodak Pro 400 PPF-120 film. A single 35-minute exposure was made from Mt Pinos, California. The field of view is 18 × 18 degrees and is centred at right ascension 17 hours 40 minutes and declination –24 degrees. North is up. The image was processed in the computer.

0013.jpg

Object name: Antares and Ophiuchus Milky Way
Designation: Star fields and nebulae (black and white)
Structures: At the right (west) is the bright red star Antares (Alpha Scorpii, magnitude 1.1). Surrounding Antares are dust particles that reflect light from the star and illuminate the orange-yellow nebula IC 4606. Even further to the right near the edge of the image is the magnitude 3.1 star Sigma Scorpii and its surrounding red emission nebula Sharpless-9. Between these two stars and slightly to the south is the large magnitude 5.4 globular cluster M4. About 3 degrees north of Antares is Rho Ophiuchi and its blue reflection nebula IC 4604. This whole area is also seen in more detail in colour image 0028.jpg of the nebula section of this CD. To the east of M4 are 11 smaller globular clusters that appear as small fuzzy specs of light. These include (from west to east) NGC 6144, NGC 6235, M62, M19, NGC 6284, NGC 6287, NGC 6293, NGC 6301, NGC 6304, NGC 6316 and NGC 6342. On a clear dark summer night you can see the slender subtle dark nebulae B44 and B45 streaming from the Antares region in the west-south-west towards a tangle of dark nebulosity within the bright Ophiuchus Milky Way to the east. The largest, darkest and most southerly of these dark nebulae is the aptly named Pipe Nebula (LDN 1773 or B65, 67, 78). The stem of this "smoker's pipe" is B65 and B67, while its "bowl" is B78. Above the bowl of the pipe is the smaller S-shaped or Snake Nebula – B72 (seen in colour in image 0064.jpg of the dark nebula section of this CD). To the north-east (upper left) are the dark nebulae B268/70, B79/276 and B84.
Photographic data: A Tamron 90 mm f/2.5 Macro lens (stopped down to f/4) was used with a light red (23A) filter and 35 mm hypersensitized Kodak Technical Pan film. A single 90-minute exposure was made from Mt Pinos. The field of view is 22 × 15 degrees and is centred at right ascension 17 hours 00 minutes and declination –25 degrees. North is up.

0014.jpg

Object name: Milky Way – Scorpius to Scutum
Designation: Star fields, bright nebulae, dark nebulae (black and white)
Structures: The southern summer Milky Way (as seen from the northern hemisphere) is seen arching across the sky from southern Scorpius at the lower right to Scutum at the upper left. The bright star clouds are split along their long axis by a long dark area of inter-stellar dust called the Great Rift. Antares can be seen at the upper right with long faint dust lanes B44 and B45 extending toward Ophiuchus above the Pipe Nebula.
Photographic data: An Olympus 35 mm f/3.5 (stopped down to f/5.6) lens was used with red (25A) filter. A single 90-minute exposure was made on hypersensitized Kodak Technical Pan 35 mm film. Taken from the Texas Star Party in west Texas.

0015.jpg

Object name: M24 or the Small Sagittarius Star Cloud
Designation: Star fields and nebulae (colour)
Structures: The Small Sagittarius Star Cloud is a rich star cloud that surrounds the two dark nebulae B92 and B93. Rarely seen in the faint red glowing hydrogen that rims these dark nebulae. To the south-east (left) is the red emission nebula IC 1283/84 and two small blue reflection nebulae NGC 6589 and 6590.
Photographic data: An Astrophysics 130 mm f/6 refractor and Pentax 6 × 7 camera were used with unhypered Kodak Pro 400 PPF-120 film. A single 45-minute exposure was made while centred at right ascension 18 hours 18 minutes and declination –18.8 degrees. The image was processed in the computer.

0016.jpg

Object name: Scutum Milky Way
Designation: Star fields, nebulae and open clusters (colour)
Structures: The upper half of the image is the dark appearing Great Rift, while the lower half consists of the easterly portion of bright star clouds of the Milky Way. On the right (south) and extending to the left (north) are emission nebulae M17, M16 and Sharpless-54, respectively. About half-way between the centre of the image and its left border is a relatively bright horn-shaped star field that extends into the darker area. This structure is called the Scutum Star Cloud. Astronomer E.E. Bernard called it the "Gem of the Milky Way". At its northern base is the rich compact open cluster M11 or Wild Duck Cluster.
Photographic data: A Pentax 165 mm f/2.8 lens (stopped down to f/5.6) and Pentax 6 × 7 camera were used with unhypered Kodak Pro 400 PPF-120 film. A single 35-minute exposure was made while centred at right ascension 18 hours 30 minutes and declination –8 degrees. The field of view is 18 × 12 degrees. North is to the left and slightly up. The image was cropped and processed in the computer.

0017.jpg and 0018.jpg (labelled)

Object name: Cygnus Milky Way
Designation: Star fields, nebulae and probable black hole (colour)
Structures: Cygnus is seen from the emission nebula NGC 7000 (North America Nebula) in the north (left) through Gamma Cygnus and emission nebula IC 1318 to dark nebula LDN 857 in the south (right). The dark nebula LDN 857 ("Fish on the Platter") is buried within the rich Cygnus star cloud on the right. In its southern region (right) is the red emission nebula Sharpless-101 and black hole candidate (X-ray source) Cygnus X-1 (right). The image is centred on our galactic equator and the long axis of the photo is parallel to it.
Photographic data: A Pentax 165 mm f/2.8 lens (stopped down to f/5.6) and Pentax 6 × 7 camera were used with unhypered Kodak Pro 400 PPF-120 film. A single 40-minute exposure was made while centred at right ascension 20 hours 20 minutes and declination +38 degrees. The field of view is 22 × 18 degrees. North is to the left and up. The image was processed in the computer.

0019.jpg and 0020.jpg (labelled)

Object name: Cepheus Milky Way
Designation: Star fields and nebulae (colour)
Structures: Emission nebulae IC 1396 and Sh2-129 are at the lower centre, while the Cocoon Nebula and dark nebula B168 are at the upper left. Emission nebula Sh2-119 is at the extreme upper left. Multiple dark nebulae can also be seen.
Photographic data: A Pentax 165 mm f/2.8 lens (stopped down to f/5.6) and Pentax 6 × 7 camera were used with unhypered Kodak Pro 400 PPF-120 film. A single 40-minute exposure was made while centred at right ascension 22 hours 00 minutes and declination +55 degrees. The field of view is 22 × 18 degrees. North is down. The image was processed in the computer.

0021.jpg

Object name: Scorpius to Ara Milky Way
Designation: Star fields, bright and dark nebulae (black and white)
Structures: Our galactic equator runs diagonally through the image from upper left to lower right and is associated with extensive star clouds and several bright and dark nebulae. Emission nebula IC 4628 (the brightest object in the field) is at the upper left (north-east) and combined emission and reflection nebula NGC 6188 is at the lower right (south-west). Just below emission nebula IC 4628 is the open cluster NGC 6231 ("Little Pleiades"), followed by a triangle of stars including the false-double Zeta Scorpii. The dark nebula SL17 (Sandqvist and Lindroos) is just above the centre of the image and appears like a black inverted U.
Photographic data: A Tamron 180 mm f/2.5 lens with hydrogen-alpha filter and hypersensitized Kodak Technical Pan film (35 mm) was used. A single 60-minute exposure was centred at right ascension 16 hours 48 minutes and declination –48 degrees. The field of view is 10 × 7 degrees and north is up. Taken from Carnegie Las Campanas Observatory, Chile, in February.

0022.jpg

Object name: IC 4628, NGC 6231 and Zeta Scorpii
Designation: Star field, nebula and open cluster (colour)
Structures: Red emission nebula IC 4628 is in the north (left), followed by open cluster NGC 6231 (called by some "Little Pleiades") and stars Zeta Scorpii in the south (right).
Photographic data: An Astrophysics 130 mm f/6 refractor was used with hypersensitized Fujicolor HG 400 (35 mm) film. Two 50-minute exposures were taken from the Carnegie Las Campanas Observatory in Chile. The image is centred at right ascension 16 hours 54 minutes and declination –42 degrees. The field of view is 2.6 × 1.7 degrees and north is to the left. A composite was made and processed in the computer.

0023.jpg

Object name: NGC 6357 ("Little Orion") and NGC6334 ("Cat's Paw")
Designation: Emission nebulae (colour)
Structures: In southern Scorpius about 6 degrees below (south-west of) the naked-eye open cluster M6 is an asterism of bright stars that has the appearance of a hockey stick (seen at the lower left of this image). Centred around the "handle of the hockey stick" is the emission nebula NGC 6357. Photographically it looks like an open flower or a faded version of the Orion Nebula M42. At the upper right of the image (south) is a somewhat brighter emission nebula called the "Cat's Paw" (NGC 6334) which indeed does resemble a cat's foot and claws.
Photographic data: An Astrophysics 130 mm f/6 refractor was used with hypersensitized Kodak Pro 400 PPF 35 mm film. Two 45-minute exposures were made from Mt Pinos while centring on right ascension 17 hours

19 minutes and declination –34 degrees. The field of view is 2.6 × 1.7 degrees and north is to the left. Digital stacking and processing was done in the computer.

0024.jpg

Object name: The Cat's Paw (NGC 6334)
Designation: Emission nebula (black and white)
Structures: A complex emission nebula in Scorpius
Photographic data: A 10-inch f/4.5 Newtonian telescope was use with a red (25A) filter and hypersensitized Kodak Technical Pan 35 mm film. A single 75-minute exposure was made from Mt San Remidio, California while centring at right ascension 17 hours 24 minutes and declination –35.7 degrees. The field of view is 1.8 × 1.2 degrees and north is up.

0025.jpg

Object name: IC 2948/44
Designation: Emission nebula (colour)
Structures: IC 2948 is a 66 × 42 minutes emission nebula that is slightly larger than M8, but rarely photographed or seen in books. It surrounds open cluster IC 2944 and is approximately 7 degrees west of the Coalsack in the southern constellation of Centaurus. Several dark-brown Bok globules are seen within the nebula.
Photographic data: An Astrophysics 130 mm f/6 refractor was used with hypersensitized Fujicolor HG 400 film (35 mm). A single 60-minute exposure was made from Carnegie Las Campanas Observatory in Chile while centring at right ascension 11 hours 38 minutes and declination –63 degrees. The field of view is 1.25 × 1.0 degrees and north is up. The image was processed in the computer.

0026.jpg

Object name: Gum 17, SL4 (Sandqvist and Lindroos), Gum 15 and VdB-Ha56 in Vela
Designation: Emission and dark nebulae (black and white)
Structures: This region is found about two degrees east of the northern border of the large Vela supernova remnant in the southern constellation Vela. The brightest of the nebulae in this image is the large treetop-shaped Gum 17. This bright nebula obscures the underlying cluster CR 203. The "trunk of the tree" is the dark nebula SL4. To the north-west (upper left) is bright nebula Gum 15. This nebula has dark fissures somewhat reminiscent of the Trifid Nebula (M20) to the north. The small bright spot at the lower right is VdB-Ha56.
Photographic data: An Astrophysics 130 mm f/6 refractor (telecompressed to f/4.5) was used with hypersensitized Kodak Technical Pan 35 mm film at the Carnegie Las Campanas Observatory in Chile. A single 60-minute exposure was made while centring at right ascension 08 hours 50 minutes and declination –42.3 degrees. The field of view is 3.5 × 2.4 degrees and north is to the left.

0027.jpg

Object name: Lagoon Nebula (M8)
Designation: Emission nebula (colour)
Structures: This red emission nebula measures 50 × 40 minutes. Its illuminating star is thought to be 9-Sagittari, seen in its western half. The brightest portion of M8 is a small area called the hour glass. Several dark areas are called Bok globules which are thought to represent the site of new star formation.
Photographic data: An Astrophysics 155 mm f/7 refractor was used with hypersensitized Kodak Pro 400 PPF (35 mm) film. Two 45-minute exposures were made while centring at right ascension 18 hours 04 minutes and declination –24.3 degrees. The field of view is 1.4 × 0.9 degrees and north is up. Digital stacking and processing were done in the computer.

0028.jpg

Object name: Antares–Rho Ophiuchus complex
Designation: Emission, reflection and dark nebulae, globular clusters (tri-colour)
Structures: This is one of the most colourful and spectacular areas (photographically) in the summer night sky. In the image one can see to the right of centre and near the bottom of the image the bright red supergiant star Antares (Alpha Scorpii, magnitude 1.1). Surrounding Antares are dust particles which reflect light from the star and illuminate the large orange-yellow nebula IC 4606. To the right of Antares is the large globular cluster M4. Above and to the left of M4 is the smaller globular cluster NGC 6144. Looking still higher the orange nebulosity fades into subtle areas of blue reflection nebulosity and the two long dark nebulae B44 and B45 that extend to the east (left). Near the upper right of the image is the star Rho Ophuchi with its large area of blue reflection nebulosity IC 4604. In the right outer quarter of the image above and slightly to the right of M4 is the magnitude 3.1 star Sigma Scorpii and its adjacent red emission nebula Sharpless-9.

Photographic data: This is a tri-colour image made from a composite of three exposures on hypered Technical Pan film. I used a Tamron 300 mm f/2.8 lens and exposed through a red (25A) filter for 45 minutes, a green (58) filter for 75 minutes and a blue (47) filter for 90 minutes. All three black and white Tech Pan negatives were developed in D-19 for 6.5 minutes at 70° F. The negatives were then scanned on to a Kodak PhotoCD so that they could be processed in a computer using Adobe Photoshop software. The centre of the image is at right ascension 16 hours 30 minutes and declination –24.8 degrees. The field of view is 6.8 × 4.5 degrees and north is up. All the exposures were from Mt Pinos, California.

0029.jpg

Object name: M16 (Eagle Nebula) and Sharpless-54
Designation: Emission nebulae
Structures: Two Milky Way emission nebulae in eastern Serpens
Photographic data: An Astrophysics 130 mm f/6 refractor and Pentax 6 × 7 camera was used with unhypered Kodak Pro 400 PPF-120 film and was centred at right ascension 18 hours 18 minutes and declination –13 degrees. Two 50-minute exposures were made from Mt Pinos. The field of view is 4.6 × 3.8 degrees and north is to the right. A composite image was made in the computer.

0030.jpg

Object name: LBN 30, 32, 35, 38 nebulae in Ophiuchus
Designation: Emission nebulae (black and white)
Structures: This is a large emission nebular complex covering an area of about 5 degrees and is not charted in *Uranometria* or other amateur atlases I have seen. It is found west of the Milky Way and north of globular cluster M107 in the constellation Ophiuchus close to the Scorpius border.
Photographic data: A Tamron 180 mm f/2.5 lens with hydrogen-alpha filter was used with hypersensitized Kodak Technical Pan 35 film. The single exposure was for 70 minutes and centred at right ascension 16 hours 30 minutes and declination –10 degrees. The field of view is about 6 × 4 degrees and north is to the right. The photo was taken from Mt Pinos, California.

0031.jpg

Object name: Sharpless-264, Barnard's Loop, Horsehead and M42
Designation: Emission nebulae (black and white)
Structures: The brightest star in the image is Betelgeuse at the upper right. Just below (4 degrees west) is the large emission nebula Sharpless-264. To the left (south) is the huge Barnard's Loop Nebula spanning the upper left portion of the image. Below this (west) are the Horsehead and Great Orion (M42) nebulae.
Photographic data: A Tamron 90 mm f/2.5 macro lens (stopped down to f/4) was used with hypersensitized Kodak Technical Pan film (35 mm) and a red (25A) filter. The single 90-minute exposure was centred at right ascension 05 hours 40 minutes and declination +3 degrees. The field of view is 22 × 15 degrees and north is to the right.

0032.jpg

Object name: Barnard's Loop, Horsehead and the Great Orion Nebula (M42)
Designation: Emission nebulae (black and white)
Structures: The bright star half-way up the image on its far right side is the middle star of Orion's Belt (Epsilon Orionis). Just below and to the left is the bright Horsehead complex which engulfs the first belt star (Zeta Orionis). The Flame Nebula (NGC 2024) lies just to the east of Zeta Orionis. Near the bottom right (south) is the over-exposed Orion Nebula, M42. On the left side of the image is a long glowing arc of bright nebulosity spanning most of the image called Barnard's Loop. Faint nebulosity is also seen between this loop and the Horsehead region.
Photographic data: A Tamron 180 mm f/2.5 lens with a hydrogen-alpha filter was used with hypersensitized Kodak Technical Pan 35 mm film. The single 70-minute exposure from Mt Pinos was centred at right ascension 05 hours 45 minutes and declination –2.5 degrees. The field of view is 11 × 7.5 degrees and north is up and slightly to the right.

0033.jpg

Object name: Orion's Sword
Designation: Emission and dark nebulosity (black and white)
Structures: The "Sword of Orion" is that region from the first belt star (Zeta Orionis) to just below M42 (Great Orion Nebula). The nebulosity of this entire region is extensive. The diminutive Horsehead dark nebula, B33 (black on the image), is seen inverted and embedded in the white surrounding nebula IC 434 at the top of the

image. At the bottom left is M42. Uncharted nebulosity can be seen bridging the area between it and the Horsehead.
Photographic data: A Tamron 300 mm f/2.8 lens with a red (25A) filter was used with hypersensitized Kodak Technical Pan 35 mm film. The single 70-minute exposure from Mt Pinos was centred at right ascension 05 hours 39 minutes and declination –3.5 degrees. The field of view is 6.7 × 4.4 degrees and north is up and slightly to the right.

0034.jpg

Object name: Rosette Nebula (NGC 2246) to Cone Nebula
Designation: Emission and dark nebulae, star fields (black and white)
Structures: On the right (south) is the wreath-like Rosette Nebula and on the left (north) the Cone Nebula. Faint nebular extensions and star fields are seen in between. The Cone Nebula is actually a very small dark nebula in the brighter complex on the left. This region is in the constellation of Monoceros.
Photographic data: A Tamron 180 mm f/2.5 lens with a hydrogen-alpha filter was used with hypersensitized Kodak Technical Pan 35 mm film. The single 70-minute exposure from Mt Pinos was centred at right ascension 06 hours 33 minutes and declination +7.3 degrees. The field of view is 8.5 × 5.7 degrees and north is to the left.

0035.jpg

Object name: Cone Nebula in Monoceros
Designation: Dark nebula embedded in bright nebulosity (black and white)
Structures: The area shows the V-shaped cluster NGC 2264 which is often called the "Christmas Tree". Near its apex (on the left) is the brightest star in the field – S-Monocerotis. This is an extremely hot giant O-type star with a luminosity equal to 8,500 Suns. Near the base of the Christmas Tree (on the right) is the dark V-shaped Cone Nebula embedded in brighter glowing gas clouds. Robert Burnham described the Cone Nebula as "a great dark pinnacle wonderfully outlined against glowing nebulosity ... The whole structure forming a picture of such strangeness and splendour that it scarcely seems natural ... Here, as in the Great Orion Nebula, even the modern observer is touched by a strange sensation of having been present at the drama of creation."
Photographic data: A Celestron 14-inch SCT (telecompressed to f/6) was used with hypersensitized Kodak Technical Pan 35 mm film. The single 70-minute exposure from Mt Pinos was centred at right ascension 06 hours 42 minutes and declination +9.7 degrees. The field of view is 51 × 34 minutes and north is to the left.

0036.jpg

Object name: IC 405 (Flaming Star Nebula – AE Auriga) and IC 410
Designation: Emission nebulae (colour)
Structures: On the right (west) is the nebula IC 405. This is primarily a red emission nebula, but has blue reflection nebulosity as well in its northern region. The illuminating star is AE Auriga which in a variable O-type. At the lower left (south-east) is the wreath-shaped emission nebula IC 410.
Photographic data: An Astrophysics 130 mm f/6 refractor which was telecompressed to f/4.5 (FL = 600 mm) was used with hypersensitized Fujicolor HG 400 35 mm film. Two 50-minute exposures were made centred at right ascension 18 hours 55 minutes and declination +34 degrees. The field of view is 3.4 × 2.3 degrees and north is up. The negatives were stacked and processed in the computer.

0037.jpg

Object name: California Nebula (NGC 1499)
Designation: Emission nebula (colour)
Structures: A large emission nebula in Perseus with a shape similar to the state of California.
Photographic data: An Astrophysics 130 mm f/6 refractor and Pentax 6 × 7 camera was used with unhypered Kodak Pro 400 PPF-120 film. Two 45-minute exposures were made centred at right ascension 04 hours 01 minute and declination +36.5 degrees. The field of view is 3.4 × 2.3 degrees and north is up. The negatives were stacked and processed in the computer.

0038.jpg

Object name: North America (NGC 7000) and Pelican (IC 5067/70) Nebulae
Designation: Emission nebulae (colour)
Structures: Two bright and aptly named red emission nebulae in the Cygnus Milky Way. The North America Nebula is the larger of the two on the left (east) and the Pelican on the right (west). Between them is the "Skull Nebula" with one "eye" blue-white and the other a smaller yellow. Our galactic equator runs diagonally through the centre of the Pelican Nebula. About 2.5 degrees to the west and off the field of this image in the bright star Deneb (Alpha Cygni).

Photographic data: An Astrophysics 130 mm f/6 refractor and Pentax 6 × 7 camera was used with unhypered Kodak Pro 400 PPF-120 film. Two 45-minute exposures were made centred at right ascension 20 hours 56 minutes and declination +44.3 degrees. The field of view is 3.4 × 2.3 degrees and north is up. The negatives were stacked, processed and cropped in the computer.

0039.jpg

Object name: Cocoon Nebula (Sh2-125 and IC 5146) and B168
Designation: Emission-reflection nebula and open cluster and dark nebula (colour)
Structures: The round Cocoon Nebula is seen at the lower left (east) of the image and the long dark nebula B168 is seen extending from it across the image to the right. The Cocoon is primarily a red emission nebula (Sharpless-125) but does have blue reflection and dark components as well. Embedded in the nebula is open cluster IC 5146.
Photographic data: An Astrophysics 155 mm f/7 refractor and Pentax 6 × 7 camera was used with unhypered Kodak Pro 400 PPF-120 film. Two 45-minute exposures were made centred at right ascension 21 hours 46 minutes and declination +48 degrees in Cygnus. The field of view is 3.4 × 2.3 degrees and north is up. The negatives were stacked, processed and cropped in the computer.

0040.jpg

Object name: NGC 6820 (Sh2-86)
Designation: Emission nebula (colour)
Structures: This faint red emission nebula in Vulpecula lies 4 degrees west of the Dumbbell Nebula (M27). The nebula has a prominent rod-shaped dark component on its eastern border. Embedded in the nebula is open cluster NGC 6823.
Photographic data: An Astrophysics 155 mm f/7 refractor and 35 mm camera were used with hypersensitized Kodak Pro 400 film. Two 50-minute exposures were made centred at right ascension 19 hours 43 minutes and declination +23.2 degrees. The field of view is 1.5 × 1.0 degrees and north is to the right. The negatives were stacked, processed and cropped in the computer.

0041.jpg

Object name: IC 1318 Gamma Cygnus Nebula
Designation: Emission nebula (colour)
Structures: To the right (west) is the bright star Gamma Cygnus. The whole area is filled with emission nebulosity designated IC 1318. The prominent dark nebula LDN 889 crosses the nebula from east to west.
Photographic data: An Astrophysics 155 mm f/7 refractor and a Pentax 6 × 7 camera were used with unhypered Kodak Pro 400 PPF-120 film. Two 50-minute exposures were made centred at right ascension 20 hours 25 minutes and declination +40.2 degrees. The field of view is 2.2 × 1.5 degrees and north is up. The negatives were stacked, processed and cropped in the computer.

0042.jpg

Object name: South-eastern IC 1318
Designation: Emission nebula (colour)
Structures: This is an enlargement of the south-eastern portion of IC 1318
Photographic data: An Astrophysics 155 mm f/7 refractor and a Pentax 6 × 7 camera were used with unhypered Kodak Pro 400 PPF-120 film. Two 50-minute exposures were made. The negatives were stacked and processed in the computer. The image was then cropped and enlarged to centre at right ascension 20 hours 28 minutes and declination +40 degrees. The field of view is 30 × 30 arc minutes and north is up.

0043.jpg

Object name: IC 1396
Designation: Emission nebula with dark components (colour)
Structures: This large red emission nebula in Cepheus has several dark components of obscuring dust (molecular clouds). In the north (right) is B161 and in the south (left) the larger B160. Bright nebula vdB142 can be seen just below (west) the centre portion of IC 1396. The bright star Mu Cephei is seen along the upper right border of the nebula. It is called the "Garnet Star" because of it ruddy colour.
Photographic data: An Astrophysics 130 mm f/6 refractor and a Pentax 6 × 7 camera were used with unhypered Kodak Pro 400 PPF-120 film. A single 50-minute exposure was made centred at right ascension 21 hours 40 minutes and declination +57.5 degrees. The field of view is 4.6 × 3.7 degrees and north is to the right. The negative was processed in the computer.

0044.jpg

Object name: IC 1805 and IC 1848
Designation: Two emission nebulae (black and white)
Structures: These two emission nebulae can be found in Cassiopeia about 5 degrees north-east of the Double Cluster of Perseus. At the lower left (south-east) is the "foetus-shaped" nebula IC 1848, while to the upper right (north-west) is "heart-shaped" IC 1805. About 45 minutes south of the southern border of IC 1805 are the poorly seen galaxies Maffei I and II.
Photographic data: A Tamron 300 mm f/2.8 lens with a red (25A) filter was used with hypersensitized Kodak Technical Pan 35 mm film. Two 60-minute exposures were made from Mt Pinos while centring at right ascension 02 hours 43 minutes and declination +60.8 degrees. The field of view is 6.8 × 4.5 degrees and north is up. The negatives were stacked and processed in the computer.

0045.jpg

Object name: IC 1805 (Sharpless-190)
Designation: Emission nebula (colour)
Structures: This emission nebula in Cassiopeia spans a distances of 2.2 degrees. It has a heart shape with a spherical extension at the upper right (north-west) named NGC 896 or IC 1795. At the centre of the nebula is open cluster Melotte-15.
Photographic data: An Astrophysics 155 mm f/7 refractor and Pentax 6 × 7 camera were used with unhypered Kodak Pro 400 PPF-120 film. A single 50-minute exposure was made while centring at right ascension 02 hours 32 minutes and declination +61.5 degrees. The field of view is 2.5 × 1.7 degrees and north is up. The negative was processed in the computer.

0046.jpg

Object name: IC 1848 (LBN 667)
Designation: Emission nebula (colour)
Structures: The emission nebula LBN 667 is often referred to as IC 1848 which represents the open star cluster buried within the nebula. This nebula lies about 2.5 degrees east of IC 1805. It has the appearance of a foetus floating on its back with its head to the upper left. Its "crown–rump" length measures 1.6 degrees.
Photographic data: An Astrophysics 155 mm f/7 refractor and a Pentax 6 × 7 camera were used with unhypered Kodak Pro 400 PPF-120 film. Two 50-minute exposures were made while centring at right ascension 02 hours 54 minutes and declination +60.5 degrees. The field of view is 3 × 2 degrees and north is up and slightly to the right. The negatives were stacked and processed in the computer.

0047.jpg

Object name: Sharpless-142 (NGC 7380)
Designation: Emission nebula and open cluster (colour)
Structures: This emission nebula is found in Cepheus close to the Lacerta border. Its red gas clouds appear convoluted because of intervening dark dust lanes. Buried within the nebula is the open cluster NGC 7380.
Photographic data: An Astrophysics 155 mm f/7 refractor and 35 mm camera were used with hypersensitized Kodak Pro 400 film. Two 50-minute exposures were made centred at right ascension 22 hours 47 minutes and declination +58 degrees. The field of view is 1.5 × 1.0 degrees and north is up. The negatives were stacked, processed and cropped in the computer.

0048.jpg

Object name: Sharpless-205
Designation: Emission nebula (black and white)
Structures: This extremely faint emission nebula crosses the Camelopardalis–Perseus border at right ascension 3 hours and 53 minutes. It is also seen to lie just south of the galactic equator at 148 degrees. The nebula is "peanut-shaped" measuring about 1.5 degrees in north–south length.
Photographic data: A Tamron 300 mm f/2.8 lens with a red (25A) filter was used with hypersensitized Kodak Technical Pan 35 mm film. A single 70-minute exposure was made from Mt Pinos while centring at right ascension 03 hours 53 minutes and declination +53 degrees. The field of view is 6 × 4 degrees and north is up.

0049.jpg

Object name: Sharpless-184 or NGC 281 ("Pac-Man Nebula")
Designation: Emission nebula (colour)
Structures: This emission nebula in Cassiopeia has a small central dark area, an eccentric open cluster (IC 1590, NGC 281) and an unusual shape that gave it its nickname.
Photographic data: An Astrophysics 155 mm f/7 refractor and 35 mm camera were used with hypersensitized Fujicolor HG 400 film. Two 50-minute exposures were made centred at right ascension 00 hours 53 minutes and declination +56.6 degrees. The field of view is 1.2 × 0.8 degrees and north is to the right. The negatives were stacked, processed and cropped in the computer

0050.jpg

Object name: Cave Nebula (Sharpless-155, LBN 529)
Designation: Emission nebula (colour)
Structures: A dim emission nebula in Cepheus. It is the concave border on the upper left of the image that gives this nebula its descriptive nickname.
Photographic data: An Astrophysics 155 mm f/7 refractor and 35 mm camera were used with hypersensitized Kodak Pro 400 PPF film. Two 50-minute exposures were made centred at right ascension 22 hours 57 minutes and declination +62.5 degrees. The field of view is 1.2 × 1.0 degrees and north is up. The negatives were stacked, processed and cropped in the computer.

0051.jpg

Object name: Cederblad-214 (Sharpless-187) and NGC 7822 (Cederblad-215)
Designation: Two emission nebulae (black and white)
Structures: On the left half of the image is NGC 7822 emission nebula, while on the right half is Cederblad-214. Cederblad-214 is the more complex of the two as it does have multiple dark nebulae extending through it as well as minimal reflecting material for this predominately emission (red) type of nebula.
Photographic data: A Tamron 300 mm f/2.8 lens with a red (25A) filter was used with hypersensitized Kodak Technical Pan 35 mm film. A single 60-minute exposure was made from Mt Pinos while centring at right ascension 00 hours 02 minutes and declination +67.8 degrees. The field of view is 4 × 2.7 degrees and north is to the left.

0052.jpg

Object name: Eta Carina Nebula (Key-hole Nebula, NGC 3372)
Designation: Large emission nebula in Carina (colour)
Structures: This is one of the most impressive objects in the sky of any type. It is a southern object in the constellation of Carina that spans about 2 degrees. The red glowing gas clouds are split by huge dark lanes giving the appearance of a starfish. The Eta Carina star which illuminates the nebula is an unstable blue supergiant a million times more luminous than our Sun.
Photographic data: An Astrophysics 130 mm f/6 refractor (telecompressed to f/4.5) and 35 mm camera were used with hypersensitized Fujicolor HG 400 film and was centred at right ascension 10 hours 44 minutes and declination –59.5 degrees. Two 50-minute exposures were made from the Carnegie Las Campanas Observatory in Chile. The field of view is 3.4 × 2.2 degrees and north is to the left. A composite image was made in the computer.

0053.jpg

Object name: IC 4628 (Gum 56)
Designation: Emission nebula (colour)
Structures: A red emission nebula in Scorpius not far north of open cluster NGC 6231. It measures 50 minutes in total length and has significant dark nebulosity in its northern half.
Photographic data: An Astrophysics 130 mm f/6 refractor and 35 mm camera were used with hypersensitized Fujicolor HG 400 film and was centred at right ascension 16 hours 57 minutes and declination –40.4 degrees. Two 60-minute exposures were made from the Carnegie Las Campanas Observatory in Chile. The field of view is 1.5 × 1degree and north is up. A composite image was made in the computer.

0054.jpg

Object name: NGC 6188 (RCW 108)
Designation: Emission nebula with reflective and dark components (colour)
Structures: In the southern constellation of Ara is this complex emission nebula with smaller reflective component at the upper right (west-south-west). Deep dark dust lanes are also seen penetrating the nebula. The open cluster NGC 6193 is below and to the right (east-south-east) of centre.

Photographic data: An Astrophysics 130 mm f/6 refractor and 35 mm camera were used with hypersensitized Fujicolor HG 400 film and was centred at right ascension 16 hours 41 minutes and declination –48.8 degrees. Two 60-minute exposures were made from the Carnegie Las Campanas Observatory in Chile. The field of view is 1.0 × 0.8 degrees and north is to the left. A composite image was made in the computer.

0055.jpg

Object name: IC 1274, IC 1275, IC 4685, NGC 6559, B303
Designation: Emission, reflection and dark nebulae (colour)
Structures: This nebular complex is found less than 1 degree east of the Lagoon Nebula (M8) in Sagittarius. It consists of a northerly (right) extension fringed in blue nebulosity (IC 1274). Just below it is the red complex of IC 1275 and IC 4685. Traversing this area from north-west to south-east is a river of dark nebulosity called B303. At the east-south-east (left) termination of B303 is a poppy-like area of red nebulosity (NGC 6559) with central blue nebulosity around a bright star.
Photographic data: An Astrophysics 155 mm f/7 refractor and 35 mm camera were used with hypersensitized Kodak Pro 400 PPF film. Two 50-minute exposures were made centred at right ascension 18 hours 09 minutes and declination –24 degrees. The field of view is 1.5 × 1.0 degrees and north is to the right. The negatives were stacked, processed and cropped in the computer.

0056.jpg

Object name: Trifid Nebula (M20, NGC 6514)
Designation: Emission and reflection nebulae (colour)
Structures: The Trifid Nebula can be found about 1 degree to the north of the Lagoon Nebula (M8). M20 consists of two nebular components. The most northerly is a blue reflection nebula that blends into the brighter southern red emission nebula. Within the southern component are the three dark fissures that give the nebula its name. The probable illuminating star appears as a double in amateur telescopes but is actually composed of six stars.
Photographic data: A Celestron 14-inch f/11 SCT and 35 mm camera were used with hypersensitized Fujicolor HG 400 film. Two 90-minute exposures were made from Mt Pinos while centring at right ascension 18 hours 02 minutes and declination –23 degrees. The field of view is 25 × 20 minutes and north is to the right. The negatives were stacked, processed and cropped in the computer.

0057.jpg

Object name: M17 (Omega or Swan Nebula) and IC 4706
Designation: Emission nebulae
Structures: An emission nebula in northern Sagittarius sometimes called the Swan because of its central bright area resemblance to that bird. To the west of M17 is the smaller and dimmer nebula IC 4706.
Photographic data: A Celestron 14-inch SCT (focally reduced to f/6) and 35 mm camera were used with hypersensitized Fujicolor HG 400 film. Two 60-minute exposures were made from Mt Pinos while centring at right ascension 18 hours 35 minutes and declination –16 degrees. The field of view is 1.2 × 1.0 degrees and north is up. The negatives were stacked, processed and cropped in the computer.

0058.jpg

Object name: M16 (Eagle Nebula)
Designation: Emission nebula (colour)
Structures: This is a bright emission nebula found in the constellation Serpens Cauda. Multiple pillars and dark molecular clouds are the sites of stellar birth.
Photographic data: A Celestron 14-inch SCT (focally reduced to f/6) and 35 mm camera were used with hypersensitized Fujicolor HG 400 film. Two 60-minute exposures were made from Mt Pinos while centring at right ascension 18 hours 19 minutes and declination –13.8 degrees. The field of view is 40 × 27 arc minutes and north is up and slightly to the left. The negatives were stacked, processed and cropped in the computer.

0059.jpg

Object name: Orion Nebula (M42, NGC 1976)) and NGC 1973/5 (Sharpless-279)
Designation: Emission and reflection nebulae (colour)
Structures: The Great Orion Nebula (M42) fills the right half of the image. The white area is over-exposed, but is the site of the Trapezium stars and a birthplace for future stars. To the left (north) is the blue reflection nebula NGC 1973/5. This is a complex structure with both blue and red components. Faint tendrils of nebulosity are seen bridging these two bright objects.
Photographic data: An Astrophysics 155 mm f/7 refractor and 35 mm camera were used with hypersensitized Kodak Pro 400 PPF film. Two 50-minute exposures were made centred at right ascension 05 hours 32 minutes and

declination –5.2 degrees. The field of view is 1.9 × 1.2 degrees and north is to the left. The negatives were stacked, processed and cropped in the computer.

0060.jpg

Object name: Rosette Nebula (NGC 2237-9, Sharpless-275)
Designation: Emission nebula (colour)
Structures: A wreath-like region of red glowing hydrogen gas. Several intricate clefts of dark nebulosity and Bok globules are also seen. The central cluster is NGC 2244.
Photographic data: A Takahashi 8-inch f/4 hyperbolic reflector and 35 mm camera were used with hypersensitized Fujicolor HG 400 film. Two 45-minute exposures were made centred at right ascension 06 hours 32 minutes and declination +5 degrees. The field of view is 2.5 × 2.0 degrees and north is to the right. The negatives were stacked, processed and cropped in the computer.

0061.jpg

Object name: Sharpless-101 (Sh2-101, Ced-173, LBN 168) and Cygnus X-1
Designation: Emission nebula and possible black hole (colour)
Structures: An emission nebula in Cygnus. Dark clefts are seen on its eastern side. South-west of this nebula is an X-ray source called Cygnus X-1. This is thought to be a black hole candidate and is located close to the marked star HDE 226868. Fourth-magnitude star Eta Cygni is 25 arc minutes to the west-south-west of this star and just out of the field.
Photographic data: An Astrophysics 155 mm f/7 refractor and 35 mm camera were used with hypersensitized Kodak Pro 400 PPF film. Two 50-minute exposures were made centred at right ascension 19 hours 59 minutes and declination +35.4 degrees. The field of view is 70 × 45 arc minutes and north is to the right. The negatives were stacked, processed and cropped in the computer.

0062.jpg

Object name: IC 410
Designation: Emission nebula (colour)
Structures: An emission nebula in Auriga with multiple dark components. It lies 2 degrees east of IC 405.
Photographic data: An Astrophysics 155 mm f/7 refractor and 35 mm camera were used with hypersensitized Kodak Pro 400 PPF film. Two 50-minute exposures were made centred at right ascension 05 hours 22 minutes and declination +33.5 degrees. The field of view is 1.5 × 1.0 degrees and north is down. The negatives were stacked, processed and cropped in the computer.

0063.jpg

Object name: Bowl of the Pipe (B78) and the Snake (B72)
Designation: Dark nebulae (black and white)
Structures: The Snake Nebula is at the upper right (north-west) and B78 at the lower left (south-east) of the image. To the lower right of the "Snake" are the three diminutive dark nebulae B68, B69 and B70. Below this area (to the south) is the brightest star in the field – Theta Ophiuchi.
Photographic data: A Tamron 300 mm f/2.8 lens with red (25A) filter was used with hypersensitized Kodak Technical Pan 35 mm film. A single 60-minute exposure was made while centred at right ascension 17 hours 30 minutes and declination – 24.5 degrees. The field of view is 6.4 × 4.3 degrees and north is up.

0064.jpg

Object name: B72 (the Snake)
Designation: Dark nebula (colour)
Structures: The Snake (B72 or LDN 66) is a dark S-shaped nebula (molecular cloud) at right ascension 17 hours 23.5 minutes and declination –23.6 degrees. To the right of it are three smaller dark nebulae B68, B69 and B70. The closest bright star to the Snake is 6.7-magnitude SAO 185357 which is seen at the four o'clock position close to B72. About 0.5 degrees below and slightly to the right of this star is a small dark nebula, B74. At the lower left corner of the image is part of B78 (the bowl of the Pipe Nebula). The brightest star in the field is 4.3-magnitude 44-Ophiuchus (SAO 185401) at the lower centre.
Photographic data: An Astrophysics 155 mm f/7 refractor and Pentax 6 × 7 camera were used for this colour image. Two 50-minute exposures were made with non-hypered Kodak Pro 400 PPF-120 film centred at right ascension 17 hours 24 minutes and declination – 24 degrees. The field of view is 2.8 × 4.2 degrees. North is up and to the left. A composite of the two negatives was made and processed in the computer.

0065.jpg

Object name: B268/70, B79/B276 and B84 in Ophiuchus
Designation: Dark nebulae (black and white)
Structures: In this image of the Ophiuchus Milky Way (north-east of the Pipe and Snake nebulae) are seen three dark nebulae. On the right (west) of the image is the largest dark nebula on the image which is designated B268/270. It does appear to resemble the Disney cartoon character Goofy in profile. At the upper centre is dark nebula B79/B276 which appears like a palm tree. To the left is the amorphous B84. All these dark nebulae represent molecular clouds that block the starlight from reaching us.
Photographic data: A Tamron 300 mm f/2.8 lens with red (25A) filter was used with hypersensitized Kodak Technical Pan 35 mm film. A single 60-minute exposure was made while centred at right ascension 17 hours 38 minutes and declination −20.8 degrees. The field of view is 6 × 4 degrees and north is up.

0066.jpg

Object name: B79 and B276 (the "Palm Tree")
Designation: Dark nebula (black and white)
Structures: This dark nebula resembles a palm tree. It spans 1 degree of sky.
Photographic data: An 8-inch Takahashi Epsilon 200 reflecting telescope was used with hypersensitized Kodak Technical Pan 35 mm film. A single 45-minute exposure was made while centred at right ascension 17 hours 38 minutes and declination −19.6 degrees. The field of view is 2 × 1.3 degrees and north is to the right.

0067.jpg

Object name: B92 and B93
Designation: Dark nebulae (black and white)
Structures: Two dark nebulae in the Small Sagittarius Star Cloud (M24). The larger of the two is B92 on the right (west).
Photographic data: An 8-inch Takahashi Epsilon 200 reflecting telescope was used with hypersensitized Kodak Technical Pan 35 mm film. A single 45-minute exposure was made while centred at right ascension 18 hours 16 minutes and declination −18.3 degrees. The field of view is 1.5 × 1.0 degrees and north is up.

0068.jpg

Object name: B142 and B143
Designation: Dark nebulae (black and white)
Structures: Two dark nebulae in Aquila about 3 degrees north-west of the bright star Altair and spanning nearly 1.5 degrees.
Photographic data: An 8-inch Takahashi Epsilon 200 reflecting telescope was used with hypersensitized Kodak Technical Pan 35 mm film. A single 45-minute exposure was made while centred at right ascension 19 hours 42 minutes and declination +11.3 degrees. The field of view is 3 × 2 degrees and north is to the left.

0069.jpg

Object name: Horsehead Nebula (B33)
Designation: Dark nebula with adjacent emission nebula (colour)
Structures: At the centre of the image is the dark Horsehead Nebula (B33) and its surrounding red emission nebulosity (IC 434). To its left is Orion's bright first belt star – Zeta Orionis. Just below (east) this star is the tan-coloured Flame Nebula (NGC 2024). Also seen are the four reflection nebulae – IC 431, IC 432, IC 435 and NGC 2023.
Photographic data: An Astrophysics 130 mm f/6 refractor and a Pentax 6 × 7 camera were used with unhypered Kodak Pro 400 PPF-120 film. Three 50-minute exposures were made while centred at right ascension 05 hours 40 minutes and declination −2.5 degrees. The field of view is 2.5 × 1.7 degrees and north is to the right. Digital stacking of the three negatives and image processing was done in the computer.

0070.jpg

Object name:
"Fish on the Platter" (LDN 857), B145, Sharpless-101 and Cygnus X-1
Designation: Dark nebulae, star fields, emission nebula and possible black hole (colour)
Structures: This beautiful and astrophysically intriguing area is rarely photographed. It is a dark nebula embedded in the Cygnus star cloud and lies just west of the Great Rift. At the top of the image (north) is the triangular dark nebula B145. The red emission nebula at the lower right (south-west) is Sharpless-101 and further to the south-west

is the bright star 21-Cygni. Between these two, lies the X-ray source and black hole candidate Cygnus X-1. Numerous areas of red glowing hydrogen intersect the northern half of the dark nebula.
Photographic data: An Astrophysics 130 mm f/6 refractor and a Pentax 6 × 7 camera were used with unhypered Kodak Pro 400 PPF-120 film. A single 50-minute exposure was made while centred at right ascension 20 hours 03 minutes and declination +35.8 degrees. The field of view is 4.5 × 3.6 degrees and north is up. Image processing was done in the computer.

0071.jpg

Object name: Double Cluster (NGC 869 and NGC 884) in Perseus
Designation: Open cluster (colour)
Structures: A young open (galactic) cluster in the constellation Perseus. The magnitudes are 5.3 and 6.1. These are two impressive clusters with a total of over 1,000 stars brighter than 12th magnitude. The ten brightest stars are young spectral types A and B supergiants. These bright stars each have a luminosity of 60,000 times that of our Sun. There are also a few M-type red supergiants which add colour to the cluster's spectacular appearance. The two clusters are of slightly different distances from Earth.
Photographic data: An Astrophysics 155 mm f/7 refractor and 35 mm camera were used with hypersensitized Kodak Pro 400 PPF film. A single 30-minute exposure was made while centring at right ascension 02 hours 20 minutes and declination +57.1 degrees. The cluster size is 30 minutes + 30 minutes = 1 degree and its brightness is 5.3 and 6.1 magnitude. The image was processed in the computer.

0072.jpg

Object name: The Pleiades (M45)
Designation: Open cluster with reflection nebulosity (colour)
Structures: The Pleiades (M45, "Seven Sisters", Subaru in Japan) is a young bright open (galactic) cluster with reflection nebulosity in Taurus. Its distance from Earth is only 400 light years and its estimated age is 20 million years. M45 is probably one of the most impressive, most frequently observed and most scientifically studied clusters in the sky. It is a cluster of several hundred stars, but with nine of major brightness. These bright stars are all young giants of spectral type B. Alcyone (Eta Tauri), the brightest, is nearly 1,000 times more luminous and ten times greater in size than the Sun. There is blue reflection nebulosity caused by the scattering of light from dust particles around the brighter stars. This is most prominent (NGC 1435) around the southern bright star Merope. Approximately 20 minutes west (four o'clock position) of the west-south-west bright star Electra (17-Tauri) is 17.9-magnitude, 1.6 × 0.2 minutes edge-on Sc spiral galaxy UGC 2838.
Photographic data: An Astrophysics 155 mm f/7 refractor and Pentax 6 × 7 camera were used with hypersensitized Kodak Pro 400 PPF 120 film. Two 45-minute exposures were made while centring at right ascension 03 hours 47 minutes and declination +24.2 degrees. The cluster's diameter is 110 minutes and its brightness is 1.2 magnitude (combined). North is up. The images were digitally stacked and processed in the computer.

0073.jpg

Object name: Berkeley-17
Designation: Oldest known open (galactic) cluster (colour)
Structures: The cluster is found about 3 degrees south of emission nebula IC 410 and 7 degrees west of SNR Simeis-147 in the constellation Auriga. It is about 13 minutes in diameter and is very faint. I have never seen a colour image of this ancient cluster and this is probably the first to appear in an amateur publication. Its age is estimated at 12.6 billion years, making it by far the oldest open cluster yet seen. NGC 6791 (0074.jpg) is the second- or third-oldest known open cluster at 9.5 billion years.
Photographic data: An Astrophysics 155 mm f/7 refractor and 35 mm camera were used with hypersensitized Kodak Pro 400 PPF. A single 40-minute exposure was made while centring at right ascension 05 hours 20.5 minutes and declination +30.5 degrees. The cluster size is 13 minutes in diameter. The image was processed in the computer.

0074.jpg

Object name: NGC 6791
Designation: Ancient open cluster (colour)
Structures: This cluster in Lyra is one of the oldest (second or third to Berkeley-17) known open clusters at an estimated age of 9.5 billion years. It is, however, much more impressive photographically than the other ancient clusters. It has a very high concentration of golden stars similar to a loose globular cluster.
Photographic data: An Astrophysics 155 mm f/7 refractor and 35 mm camera were used with hypersensitized Kodak Pro 400 PPF film. A single 45-minute exposure was made while centring at right ascension 19 hours 21 minutes and declination +37.8 degrees. The cluster size is 15 minutes in diameter. The image was processed in the computer.

0075.jpg

Object name: NGC 188
Designation: Ancient open cluster (colour)
Structures: NGC 188 is yet another very old open cluster at 7.2 billion years old. This cluster is looser than NGC 6791. It is quite close to the north celestial pole and only 4 degrees from Polaris.
Photographic data: An Astrophysics 155 mm f/7 refractor and 35 mm camera were used with hypersensitized Kodak Pro 400 PPF film. Two 40-minute exposures were made while centring at right ascension 00 hours 47.5 minutes and declination +85.2 degrees. The cluster size is 13 minutes in diameter and its brightness is 8.1 magnitude. The image was digitally stacked and processed in the computer.

0076.jpg

Object name: M6 (NGC 6405)
Designation: Open cluster (colour)
Structures: This is a bright open cluster (magnitude 4.2) about 4 degrees north-west of M7. Most of the brightest stars are white or blue-white of spectral class B or A, but there is also a conspicuous star (BM Scorpii) with a bright red-orange colour at the four o'clock position (north is up).
Photographic data: An Astrophysics 155 mm f/7 refractor and 35 mm camera were used with hypersensitized Kodak Pro 400 PPF film. A single 30-minute exposure was made while centring at right ascension 17 hours 40 minutes and declination –32.6 degrees. The cluster size is 26 minutes and its brightness is 4.2 magnitude. The image was processed in the computer.

0077.jpg

Object name: M7 (NGC 6475)
Designation: Open cluster (colour)
Structures: This is a bright cluster (magnitude 3.3) with 80 stars brighter than magnitude 10. It is almost 1 degree across. It has an hour-glass appearance and is found in the constellation Scorpius.
Photographic data: An Astrophysics 155 mm f/7 refractor and 35 mm camera were used with hypersensitized Kodak Pro 400 PPF film. A single 30-minute exposure was made while centring at right ascension 17 hours 53 minutes and declination –34.8 degrees. The cluster size is 80 minutes and its brightness is 3.3 magnitude. The image was processed in the computer.

0078.jpg

Object name: M46 (NGC 2437)
Designation: Open cluster with planetary nebula (colour)
Structures: This is a fairly rich cluster in the constellation Puppis. The red ring-like structure at six o'clock (north) is planetary nebula NGC 2438. This planetary nebula is not thought to be part of the cluster and is probably a foreground object.
Photographic data: An Astrophysics 155 mm f/7 refractor and 35 mm camera were used with hypersensitized Kodak Pro 400 PPF film. Three 30-minute exposures were made while centring at right ascension 07 hours 39 minutes and declination –14.6 degrees. The cluster's size is 27 minutes and its brightness is 6.1 magnitude. The three negatives were digitally stacked and processed in the computer.

0079.jpg

Object name: Tombaugh-4
Designation: Open cluster (colour)
Structures: This is a small open cluster located in the north-west portion of the emission nebula IC 1805 in the constellation Cassiopeia. It is not charted in my atlases and I found it accidentally when I photographed the nebula where it is located. It was discovered by Clyde Tombaugh who also discovered the planet Pluto.
Photographic data: An Astrophysics 155 mm f/7 refractor and Pentax 6 × 7 camera were used with Kodak Pro 400 PPF 120 film. A single 50-minute exposure was made while centring at right ascension 02 hours 28 minutes and declination +61.8 degrees. The image was processed and enlarged in the computer

0080.jpg

Object name: NGC 7789
Designation: Open cluster (colour)
Structures: This is an unusually rich galactic star cluster in Cassiopeia with about 300 stars in a 15 arc minute circle.
Photographic data: An Astrophysics 155 mm f/7 refractor and 35 mm camera were used with hypersensitized Kodak Pro 400 PPF film. A single 40-minute exposure was made while centring at right ascension 23 hours

57 minutes and declination +56.6 degrees. The cluster is 15 minutes in diameter and its brightness is 6.7 magnitude. The image was processed in the computer.

0081.jpg

Object name: NGC 457 ("Owl Cluster")
Designation: Open cluster (colour)
Structures: A remarkably bright (magnitude 6.4) cluster in Cassiopeia. The cluster does look like an owl with two asymmetrical bright eyes.
Photographic data: An Astrophysics 155 mm f/7 refractor and 35 mm camera were used with hypersensitized Kodak Pro 400 PPF film. A single 40-minute exposure was made while centring at right ascension 01 hour 19.5 minutes and declination +58.2 degrees. The cluster is 13 minutes in size and its brightness is 6.4 magnitude. The image was processed in the computer.

0082.jpg

Object name: M38 (NGC1912) and NGC 1907
Designation: Open cluster (colour)
Structures: This open cluster in Auriga is about 4,200 light years from Earth and contains some 100 stars. The brightest stars are of spectral type G, A and B. The smaller open cluster NGC 1907 is seen to the right.
Photographic data: An Astrophysics 155 mm f/7 refractor and 35 mm camera were used with hypersensitized Kodak Pro 400 PPF film. A single 40-minute exposure was made while centring at right ascension 05 hours 28.7 minutes and declination +35.8 degrees. M38 is 21 minutes in size and its brightness is 6.4 magnitude. The image was processed in the computer.

0083.jpg

Object name: NGC 6939 (open cluster) and NGC 6946 (spiral galaxy)
Designation: Open cluster and galaxy (colour)
Structures: The open cluster NGC 6939 is seen at the upper right and spiral galaxy NGC 6946 is seen at the lower left. The galaxy (much further from us than the open cluster) is a face-on spiral with several emission nebulae seen in its arms appearing as small red dots.
Photographic data: An Astrophysics 155 mm f/7 refractor and 35 mm camera were used with hypersensitized Kodak Pro 400 PPF film. Two 45-minute exposures were made while centring at right ascension 20 hours 33 minutes and declination +60.5 degrees. Open cluster NGC 6939 is 7 minutes in size and its brightness is 7.8 magnitude, while galaxy NGC 6946 is 11.6 × 9.8 minutes and is 9.6 magnitude. The two negatives were digitally stacked and processed in the computer. North is up.

0084.jpg

Object name: NGC 3532
Designation: Open cluster (colour)
Structures: This is a large bright American-football-shaped open cluster in the southern constellation of Carina. It can be found less than 3 degrees west-north-west from the centre of the large Eta Carina (Keyhole) Nebula. Sir John Herschel considered it the "most brilliant he had ever seen". It is large, covering nearly a degree and containing 150 stars down to magnitude 12. Over 90% of the brighter stars are of spectral type A.
Photographic data: An Astrophysics 130 mm f/6 refractor and 35 mm camera were used with hypersensitized Fujicolor HG 400 film. Two 30-minute exposures were made from the Carnegie Las Campanas Observatory in Chile while centring at right ascension 11 hours 05 minutes and declination –58.7 degrees. The cluster is 55 minutes in size and its brightness is 3.0 magnitude. The two negatives were digitally stacked and processed in the computer.

0085.jpg

Object name: 47-Tucanae (NGC 104)
Designation: Globular cluster (colour)
Structures: This is considered the second-brightest globular cluster in the sky. It lies in the southern sky not far from the Small Magellanic Cloud.
Photographic data: An Astrophysics 155 mm f/7 refractor was used with Fujicolor G-800 film and a 35 mm camera. A single 20-minute exposure was made from the shores of Lake Titicaca, Bolivia, while centred at right ascension 00 hours 24 minutes and declination –72 degrees. The globular cluster's size is 50 minutes and its brightness is 4.0 magnitude. The image was processed in the computer

0086.jpg

Object name: Omega Centauri (NGC 5139)
Designation: Globular cluster (colour)
Structures: This is the brightest and largest appearing globular cluster in the night sky. Its total luminosity is equal to one million Suns. The globular cluster is found in the southern constellation of Centaurus. It was photographed from southern California where it only rises to a maximum of 8 degrees above the horizon and is therefore very difficult to photograph. The image seen here is degraded by poor seeing and transparency and is therefore not of optimal quality for the telescope that was used.
Photographic data: A Celeston 14-inch f/11 SCT telescope was used at f/6 with hypersensitized Fujicolor HG 400 film. A single 45-minute exposure was made from the southern California desert while centring at right ascension 13 hours 27 minutes and declination –47.5 degrees. The globular cluster's size is 53 minutes and its brightness is 3.9 magnitude. The image was processed in the computer.

0087.jpg

Object name: IC 443 (LBN 844)
Designation: Supernova remnant (SNR) (colour)
Structures: An expanding filamentary gas cloud in Gemini with moderately strong radio source. The bright star is Eta Geminorum.
Photographic data: An Astrophysics 155 mm f/7 refractor and 35 mm camera were used with hypersenstized Kodak Pro 400 PPF film. Two 50-minute exposures were made while centred at right ascension 06 hours 32 minutes and declination +22.5 degrees. The field of view is 1.2 × 0.8 degrees and north is up. Digital stacking and image processing were done in the computer.

0088.jpg

Object name: Simeis-147 (Sharpless-240)
Designation: Supernova remnant (SNR) (black and white)
Structures: This is one of the largest and faintest objects in the sky covering an area of 3 degrees (6 full moons). It is a supernova remnant not far from the Crab Nebula (M1) in Taurus. Like other SNRs, it has a complex filamentary structure and radio source. It was discovered in the early 1950s at the Crimean Astrophysical Observatory. This image shows the south-west half of the object. It is the faintest object on this CD.
Photographic data: An Astrophysics 130 mm f/6 refractor that was telecompressed to f/4.5, red (25A) filter and hypersensitized Kodak Technical Pan (35 mm) film were used. A single 2-hour (120 minutes) exposure was made from Mt Pinos while centred at right ascension 06 hours 32 minutes and declination +22.5 degrees. The field of view is 1.2 × 0.8 degrees and north is up. Image processing was done in the computer with Adobe Photoshop 4.0 software.

0089.jpg

Object name: Veil Nebula (all three components – NGC 6960, 74, 79, 92, 95)
Designation: Supernova remnant (colour)
Structures: All three components of the Veil Nebula in Cygnus. This SNR is much brighter than IC 443 and Simeis-147. The middle component is the most difficult to observe visually.
Photographic data: An Astrophysics 130 mm f/6 refractor and a Pentax 6 × 7 camera were used with unhypered Kodak Pro 400 PPF-120 film. A single 50-minute exposure was made while centred at right ascension 20 hours 51 minutes and declination +31degrees. The field of view is 4.0 × 3.2 degrees and north is up and slightly to the left. Image processing was done in the computer.

0090.jpg

Object name: Eastern Veil Nebula (NGC 6992/95)
Designation: Supernova remnant (colour)
Structures: This is the eastern portion of the Veil and its largest single component. Again note the filamentary structure common to supernova remnants with expanding gas shells.
Photographic data: An Astrophysics 155 mm f/7 refractor and 35 mm camera were used with hypersenstized Kodak Pro 400 PPF film. Two 50-minute exposures were made while centred at right ascension 20 hours 56 minutes and declination +31.5 degrees. The field of view is 1.7 × 1.2 degrees and north is to the right. Digital stacking and image processing were done in the computer.

0091.jpg

Object name: NGC 891
Designation: Galaxy (black and white)

Structures: A spiral galaxy in the constellation of Andromeda. This galaxy is seen edge-on with the nuclear bulge in the centre and the galactic equator as a dark linear profile extending the length of the disc. Most of the dust and star formation occurs in this relatively narrow space.
Photographic data: A Celestron 14-inch f/11 (focally reduced to f/6) SCT was used with a 35 mm camera and hypersensitized Kodak Technical Pan film. A single 90-minute exposure was made while centred at right ascension 02 hours 23 minutes and declination +42.3 degrees. The galaxy's size is 13.5 × 2.4 minutes and its brightness is 10.8 magnitude.

0092.jpg

Object name: M108 (galaxy) and M97 (planetary nebula)
Designation: Galaxy and planetary nebula (black and white)
Structures: In the constellation Ursa Major are two deep sky objects in the same field. On the right (south) of this image is the Owl Nebula (M97, NGC 3587). This is a planetary nebula with two dark areas suggesting the eyes of an owl. On the left (north) and much further away is the spiral galaxy M108 (NGC 3556).
Photographic data: An 8-inch Takahashi Epsilon 200 reflecting telescope was used with hypersensitized Kodak Technical Pan 35 mm film. A single 45-minute exposure was made while centred at right ascension 11 hours 12 minutes and declination +55.3 degrees. The field of view is 1.5 × 1.0 degrees and north is to the left. M97 is 2.8 minutes in diameter and its brightness is 12 magnitude. M108's size is 8.7 × 2.2 minutes and its brightness is 10.7 magnitude.

0093.jpg

Object name: Virgo Cluster (Markarian's Chain)
Designation: Galaxy group (black and white)
Structures: This is only a portion of the Virgo Cluster of galaxies. The galaxies shown here are in the constellation Virgo, but are also very close to the constellation Coma Berenices to the north. On the right (west) side of the image is the elliptical galaxy M84. Close to the centre is the brightest object in the field – M86 (magnitude 9.8) which is also an elliptical galaxy. On the left side of the image is the spiral galaxy NGC 4438 which has its disc distorted from its close neighbour to the north (above) – NGC 4435. Several other galaxies are also seen in the image.
Photographic data: An Astrophysics 155 mm f/7 refractor and 35 mm camera were used with hypersensitized Kodak Technical Pan 35 mm film. A single 60-minute exposure was made centred at right ascension 12 hours 26 minutes and declination +12.7 degrees. The field of view is 1.2 × 0.8 degrees and north is up. M86 is the brightest and largest appearing object in the field. Its size is 6.9 × 5.7 minutes and its brightness is 9.8 magnitude. The negative was processed in the computer.

0094.jpg

Object name: The Trio in Leo – M65, M66 and NGC 3628
Designation: Galaxies (colour)
Structures: In the constellation of Leo are three bright closely spaced galaxies called the "Trio in Leo". Just to the right of the centre of the image is the spiral galaxy M66. Above centre is a more symmetrical spiral galaxy – M65. At the lower left is the nearly edge-on spiral galaxy NGC 3628.
Photographic data: An Astrophysics 155 mm f/7 refractor and 35 mm camera were used with hypersensitized Kodak Pro 400 PPF 35 mm film. A single 45-minute exposure was made centred at right ascension 11 hours 19 minutes and declination +13.2 degrees. The field of view is 1.2 × 0.8 degrees and north is to the left. NGC 3628 is the largest appearing galaxy in the image. Its size is 14.8 × 2.9 minutes and its brightness is 10.3 magnitude. The negative was processed in the computer.

0095.jpg

Object name: NGC 253 (galaxy) and NGC 288 (globular cluster)
Designation: Galaxy and globular cluster (colour)
Structures: In the constellation Sculptor are two structures that appear close together but are of greatly different distance. In the halo around our Galaxy is the globular cluster NGC 288 seen on the left (south) of this image. At a much greater distance is the large spiral galaxy NGC 253 (27.7 × 6.7 minutes) seen on the right.
Photographic data: An Astrophysics 155 mm f/7 refractor and Pentax 6 × 7 camera were used with unhypered Kodak Pro 400 PPF-120 film. A single 45-minute exposure was made centred at right ascension 00 hours 50 minutes and declination –26 degrees. The field of view is 2.8 × 1.8 degrees and north is up. NGC 253 size is 27.7 × 6.7 minutes and its brightness is 8.0 magnitude. NGC 288 is 13 minutes in diameter and its brightness is 8.1 magnitude. The negative was processed in the computer.

0096.jpg

Object name: Whirlpool Galaxy (M51, NGC 5194) and Supernova SN1994i
Designation: Galaxy with exploding supernova (colour)
Structures: The beautiful spiral Whirlpool Galaxy had an actively exploding supernova in April 1994. The supernova is seen at about the 11 o'clock position relative to the galaxy's nucleus.
Photographic data: A Celestron 14-inch f/11 SCT was used with a 35 mm camera and hypersensitized Kodak Fujicolor HG 400 film. A single 90-minute exposure was made while centred at right ascension 13 hours 30 minutes and declination +47.2 degrees. The galaxy's size is 11.3 × 6.9 minutes and its brightness is 9.0 magnitude. The exposure was made on the night of 16 April 1994. The negative was processed in the computer.

0097.jpg

Object name: M33 (NGC 598)
Designation: Galaxy (colour)
Structures: This relatively close galaxy in Triangulum is a member of the Local Group of Galaxies. Its brighter members include our Milky Way, the Magellanic Clouds, M31 and M33. It is classified as an Sc spiral and its distance from Earth is approximately 2.3 million light years. Several of its brighter emission nebulae (H-II regions) can be seen as red spots in its spiral arms.
Photographic data: An Astrophysics 155 mm f/7 refractor and Pentax 6 × 7 camera were used with unhypered Kodak Pro 400 PPF-120 film. Two 50-minute exposures were made centred at right ascension 01 hour 34 minutes and declination +30.6 degrees. The field of view is 2.8 × 1.8 degrees. M33 is 65 × 38 minutes and its brightness is 6.3 magnitude. The negatives were digitally stacked and processed in the computer.

0098.jpg

Object name: Andromeda Galaxy (M31)
Designation: Galaxy (colour)
Structures: This is our closest full-sized neighbour galaxy. It can be seen on a dark night as a glow spanning nearly 3 degrees of sky (6 full moons). The blue spiral arms with young stars contrast to the more white-yellow nuclear area with older stars. The dark dust lanes in the spiral arms are also well seen. A focal condensation of blue stars in an outer arm at the seven o'clock position is named NGC 206. The two satellite galaxies are M32 (NGC 221) at the ten o'clock and M110 (NGC 205) at the five o'clock positions. The image is inverted from its usual position as viewed in the sky.
Photographic data: An Astrophysics 130 mm f/6 refractor and Pentax 6 × 7 camera were used. A total of four exposures were made. Two 50-minute exposures with unhypered Kodak Pro 400 PPF-120 film and another two 30-minute exposures with unhypered Kodak Pro 1000 PMZ-120 film. The telescope was centred at right ascension 00 hours 43 minutes and declination +41.2 degrees. The field of view is 3.5 × 2.3 degrees. M31 size is 192 × 62 minutes and its brightness is 4.4 magnitude. The four negatives were digitally stacked and processed in the computer. The image is inverted.

0099.jpg

Object name: Stephan's Quintet (NGC 7317, 7318A, 7318B, 7319 and 7320) and NGC 7331
Designation: Galaxies (colour)
Structures: At the upper right of the image is the spiral galaxy NGC 7331. At the lower left (30 arc minutes south) is a small cluster of five galaxies called Stephan's Quintet. They are NGC 7317, NGC 7318A, NGC 7318B, NGC 7319 and NGC 7320. The largest member of the group is NGC 7320.
Photographic data: An Astrophysics 155 mm f/7 refractor and Pentax 6 × 7 camera were used with unhypered Kodak Pro 400 PPF-120 film. Two 50-minute exposures were made centred at right ascension 22 hours 37 minutes and declination +34.2 degrees. The field of view is 1.5 × 1.0 degrees. NGC 7331 size is 14.5 × 3.7 minutes and its brightness is 9.4 magnitude. The negatives were digitally stacked and processed in the computer.

0100.jpg

Object name: UGC 2838 near M45 (Pleiades)
Designation: Galaxy (colour)
Structures: Most of the atlases that amateur astronomers use show the bright Pleiades star cluster (M45) without any nearby galaxy listed. It was a surprise to me to find the small edge-on spiral galaxy UGC 2838 in one of my earlier photos. It can be seen at the four o'clock position (west-south-west) relative to Electra (one of the nine major stars of the Pleiades). UGC 2838 is said to be a 17.9-magnitude galaxy measuring 1.6 × 0.2 arc minutes. There are actually other galaxies in this region, but they are too faint for me to record with my equipment.
Photographic data: An Astrophysics 155 mm f/7 refractor and Pentax 6 × 7 camera were used with unhypered Kodak Pro 400 PPF-120 film. Two 45-minute exposures were made centred at right ascension 03 hours 44 minutes

and declination +24.0 degrees. The field of view is 0.5 × 0.3 degrees. UGC 2838 size is 1.6 × 0.2 minutes and its brightness is 17.9 magnitude. The negatives were digitally stacked, cropped and processed in the computer.

0101.jpg

Object name: M82 (NGC 3034)
Designation: Galaxy (colour)
Structures: This is an unusual spindle-shaped galaxy that is known to have a strong radio source. Most of the galaxy has a blue colour and smooth contours. However, its central region has a bizarre multicoloured appearance with nodularity and subtle filamentary rays extending outwards toward space. M82 is in Ursa Major and is only half a degree north of its larger and more normal appearing neighbour M81.
Photographic data: A Celestron 14-inch f/11 SCT was used with a 35 mm camera and hypersensitized Fujicolor HG 400 film. A single 60-minute exposure was made while centred at right ascension 09 hours 56 minutes and declination +69.7 degrees. The galaxy's size is 11.3 × 4.2 minutes and its brightness is 9.3 magnitude. The image was processed in the computer.

0102.jpg

Object name: M64 (NGC 4826) – Black Eye Galaxy
Designation: Galaxy (colour)
Structures: This galaxy gets its name from the dark dust lane near the nucleus. It has a relatively high surface brightness requiring a shorter photographic exposure than most other galaxies.
Photographic data: A Celestron 11-inch f/10 SCT was used with a 35 mm camera and hypersensitized Kodak Pro 400 PPF film. A single 45-minute exposure was made while centred at right ascension 12 hours 56 minutes and declination +21.6 degrees. The galaxy's size is 10.1 × 5.4 minutes and its brightness is 9.4 magnitude. The image was processed in the computer.

0103.jpg

Object name: NGC 4565
Designation: Galaxy (colour)
Structures: The classic edge-on spiral galaxy in the constellation Coma Berenices. It resembles NGC 891 but appears to have a more localized nuclear bulge and a narrower profile.
Photographic data: A Celestron 11-inch f/10 SCT was used with a 35 mm camera and hypersensitized Kodak Pro 400 PPF film. A single 90-minute exposure was made while centred at right ascension 12 hours 36 minutes and declination +26 degrees. The galaxy's size is 15.9 × 1.8 minutes and its brightness is 10.4 magnitude. The image was processed in the computer.

0104.jpg

Object name: Michael A. Stecker, MD
Designation: Astrophotographer (colour)
Structures: Michael on a sunny day at 8,300 ft, Mt Pinos, California.
Photographic data: Canon A-1 camera with 50 mm f/1.4 lens and Ektachrome 100 film. Exposure was 1/250 second at f/11.

Colour Plate Section

0069.jpg

Object name: Horsehead Nebula (B33)

Designation: Dark nebula with adjacent emission nebula (colour)

Structures: At the centre of the image is the dark Horsehead Nebula (B33) and its surrounding red emission nebulosity (IC 434). To its left is the bright first belt star – Zeta Orionis. Just below (east) of this star is the tan-coloured Flame Nebula (NGC 2024). Also seen are the four reflection nebulae – IC 431, IC 432, IC 435 and NGC 2023.

Photographic data An Astrophysics 130 mm f/6 refractor and a Pentax 6 × 7 camera were used with unhypered Kodak Pro 400 PPF-120 film. Three 50-minute exposures were made while centred at right ascension 05 hours 40 minutes and declination −2.5 degrees. The field of view is 2.5 × 1.7 degrees and north is to the right. Digital stacking of the three negatives and image processing was done in the computer.

This image should be used to calibrate your monitor to view images from the CD-ROM. When viewing this image on your screen, adjust contrast and brightness until the view matches this photograph as closely as possible. Your monitor is then set up to view all the images on the CD-ROM.

0064.jpg

Object name: B72 (the Snake)
Designation: Dark nebula (colour)
Structures: The Snake (B72 or LDN 66) is a dark S-shaped nebula (molecular cloud) at right ascension 17 hours 23.5 minutes and declination –23.6 degrees. To the right of it are three smaller dark nebulae B68, B69 and B70. The closest bright star to the Snake is 6.7-magnitude SAO 185357 which is seen at the four o'clock position close to B72. About 0.5 degrees below and slightly to the right of this star is a small dark nebula, B74. At the lower left corner of the image is part of B78 (the bowl of the Pipe Nebula). The brightest star in the field is 4.3-magnitude 44-Ophiuchus (SAO 185401) at the lower centre.
Photographic data: An Astrophysics 155 mm f/7 refractor and Pentax 6 × 7 camera were used for this colour image. Two 50-minute exposures were made with non-hypered Kodak Pro 400 PPF-120 film centred at right ascension 17 hours 24 minutes and declination – 24 degrees. The field of view is 2.8 × 4.2 degrees. North is up and to the left. A composite of the two negatives was made and processed in the computer.

0039.jpg

Object name: Cocoon Nebula (Sh2-125 and IC 5146) and B168
Designation: Emission-reflection nebula and open cluster and dark nebula (colour)
Structures: The round Cocoon Nebula is seen at the lower left (east) of the image and the long dark nebula B168 is seen extending from it across the image to the right. The Cocoon is primarily a red emission nebula (Sharpless-125) but does have blue reflection and dark components as well. Embedded in the nebula is open cluster IC 5146.
Photographic data: An Astrophysics 155 mm f/7 refractor and Pentax 6 × 7 camera was used with unhypered Kodak Pro 400 PPF-120 film. Two 45-minute exposures were made centred at right ascension 21 hours 46 minutes and declination +48 degrees in Cygnus. The field of view is 3.4 × 2.3 degrees and north is up. The negatives were stacked, processed and cropped in the computer.

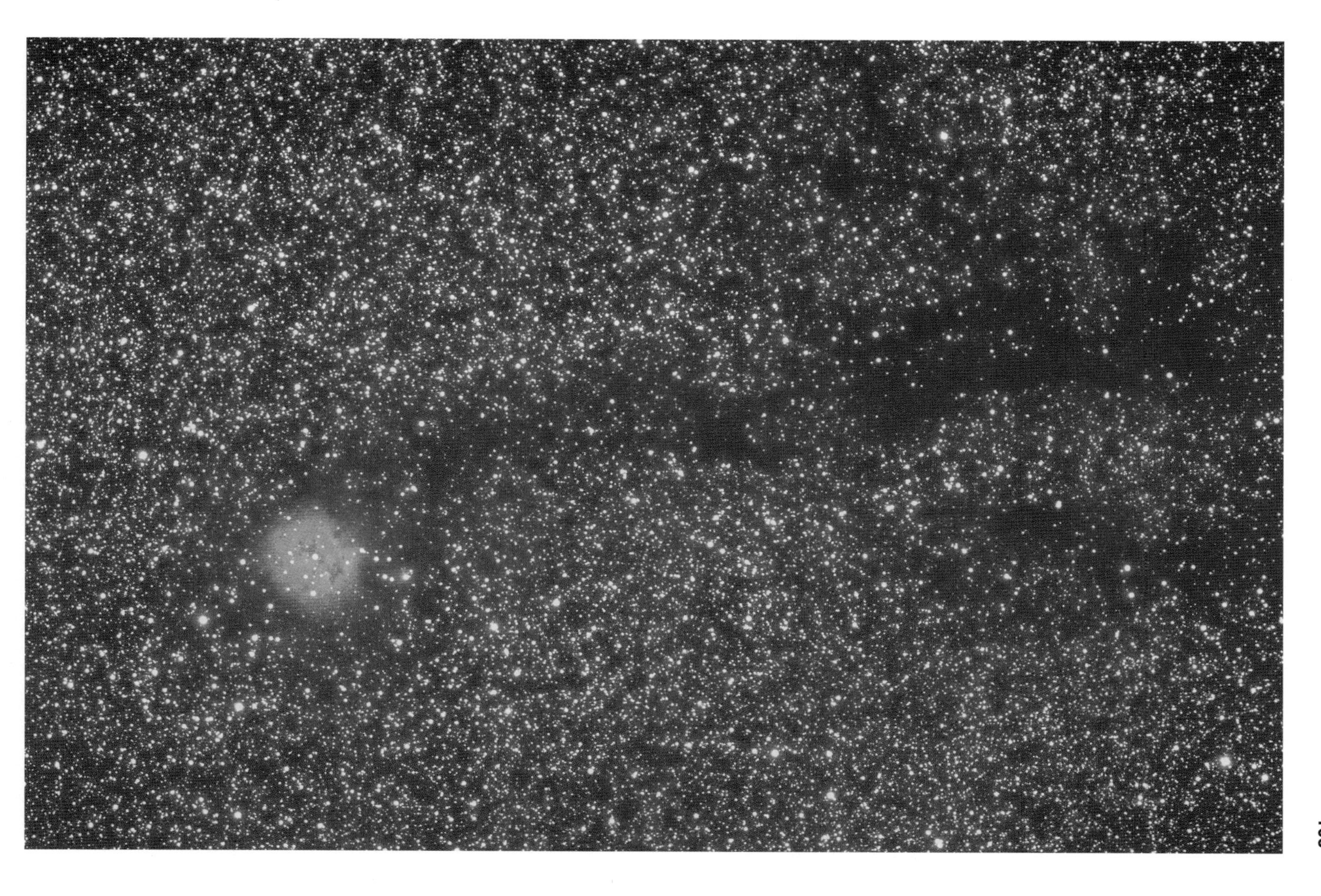

0043.jpg

Object name: IC 1396

Designation: Emission nebula with dark components (colour)

Structures: This large red emission nebula in Cepheus has several dark components of obscuring dust (molecular clouds). In the north (right) is B161 and in the south (left) the larger B160. Bright nebula vdB142 can be seen just below (west) the centre portion of IC 1396. The bright star Mu Cephei is seen along the upper right border of the nebula. It is called the "Garnet Star" because of it ruddy colour.

Photographic data: An Astrophysics 130 mm f/6 refractor and a Pentax 6 × 7 camera were used with unhypered Kodak Pro 400 PPF-120 film. A single 50-minute exposure was made centred at right ascension 21 hours 40 minutes and declination +57.5 degrees. The field of view is 4.6 × 3.7 degrees and north is to the right. The negative was processed in the computer.

0050.jpg

Object name: Cave Nebula (Sharpless-155, LBN 529)
Designation: Emission nebula (colour)
Structures: A dim emission nebula in Cepheus. It is the concave border on the upper left of the image that gives this nebula its descriptive nickname.
Photographic data: An Astrophysics 155 mm f/7 refractor and 35 mm camera were used with hypersensitized Kodak Pro 400 PPF film. Two 50-minute exposures were made centred at right ascension 22 hours 57 minutes and declination +62.5 degrees. The field of view is 1.2 × 1.0 degrees and north is up. The negatives were stacked, processed and cropped in the computer.

0028.jpg

Object name: Antares–Rho Ophiuchus complex

Designation: Emission, reflection and dark nebulae, globular clusters (tri-colour)

Structures: This is one of the most colourful and spectacular areas (photographically) in the summer night sky. In the image one can see to the right of centre and near the bottom of the image the bright red supergiant star Antares (Alpha Scorpii, magnitude 1.1). Surrounding Antares are dust particles which reflect light from the star and illuminate the large orange-yellow nebula IC 4606. To the right of Antares is the large globular cluster M4. Above and to the left of M4 is the smaller globular cluster NGC 6144. Looking still higher the orange nebulosity fades into subtle areas of blue reflection nebulosity and the two long dark nebulae B44 and B45 that extend to the east (left). Near the upper right of the image is the star Rho Ophuchi with its large area of blue reflection nebulosity IC 4604. In the right outer quarter of the image above and slightly to the right of M4 is the magnitude 3.1 star Sigma Scorpii and its adjacent red emission nebula Sharpless-9.

Photographic data: This is a tri-colour image made from a composite of three exposures on hypered Technical Pan film. I used a Tamron 300 mm f/2.8 lens and exposed through a red (25A) filter for 45 minutes, a green (58) filter for 75 minutes and a blue (47) filter for 90 minutes. All three black and white Tech Pan negatives were developed in D-19 for 6.5 minutes at 70° F. The negatives were then scanned on to a Kodak PhotoCD so that they could be processed in a computer using Adobe Photoshop software. The centre of the image is at right ascension 16 hours 30 minutes and declination –24.8 degrees. The field of view is 6.8×4.5 degrees and north is up. All the exposures were from Mt Pinos, California.

0072.jpg

Object name: The Pleiades (M45)

Designation: Open cluster with reflection nebulosity (colour)

Structures: The Pleiades (M45, "Seven Sisters", Subaru in Japan) is a young bright open (galactic) cluster with reflection nebulosity in Taurus. Its distance from Earth is only 400 light years and its estimated age is 20 million years. M45 is probably one of the most impressive, most frequently observed and most scientifically studied clusters in the sky. It is a cluster of several hundred stars, but with nine of major brightness. These bright stars are all young giants of spectral type B. Alcyone (Eta Tauri), the brightest, is nearly 1,000 times more luminous and ten times greater in size than the Sun. There is blue reflection nebulosity caused by the scattering of light from dust particles around the brighter stars. This is most prominent (NGC 1435) around the southern bright star Merope. Approximately 20 minutes west (four o'clock position) of the west-south-west bright star Electra (17-Tauri) is 17.9-magnitude, 1.6 × 0.2 minutes edge-on Sc spiral galaxy UGC 2838.

Photographic data: An Astrophysics 155 mm f/7 refractor and Pentax 6 × 7 camera were used with hypersensitized Kodak Pro 400 PPF 120 film. Two 45-minute exposures were made while centring at right ascension 03 hours 47 minutes and declination +24.2 degrees. The cluster's diameter is 110 minutes and its brightness is 1.2 magnitude (combined). North is up. The images were digitally stacked and processed in the computer.

0086.jpg

Object name: Omega Centauri (NGC 5139)
Designation: Globular cluster (colour)
Structures: This is the brightest and largest appearing globular cluster in the night sky. Its total luminosity is equal to one million Suns. The globular cluster is found in the southern constellation of Centaurus. It was photographed from southern California where it only rises to a maximum of 8 degrees above the horizon and is therefore very difficult to photograph. The image seen here is degraded by poor seeing and transparency and is therefore not of optimal quality for the telescope that was used.
Photographic data: A Celeston 14-inch f/11 SCT telescope was used at f/6 with hypersensitized Fujicolor HG 400 film. A single 45-minute exposure was made from the southern California desert while centring at right ascension 13 hours 27 minutes and declination –47.5 degrees. The globular cluster's size is 53 minutes and its brightness is 3.9 magnitude. The image was processed in the computer.

0089.jpg

Object name: Veil Nebula (all three components – NGC 6960, 74, 79, 92, 95)
Designation: Supernova remnant (colour)
Structures: All three components of the Veil Nebula in Cygnus. This SNR is much brighter than IC 443 and Simeis-147. The middle component is the most difficult to observe visually.
Photographic data: An Astrophysics 130 mm f/6 refractor and a Pentax 6 × 7 camera were used with unhypered Kodak Pro 400 PPF-120 film. A single 50-minute exposure was made while centred at right ascension 20 hours 51 minutes and declination +31degrees. The field of view is 4.0 × 3.2 degrees and north is up and slightly to the left. Image processing was done in the computer.

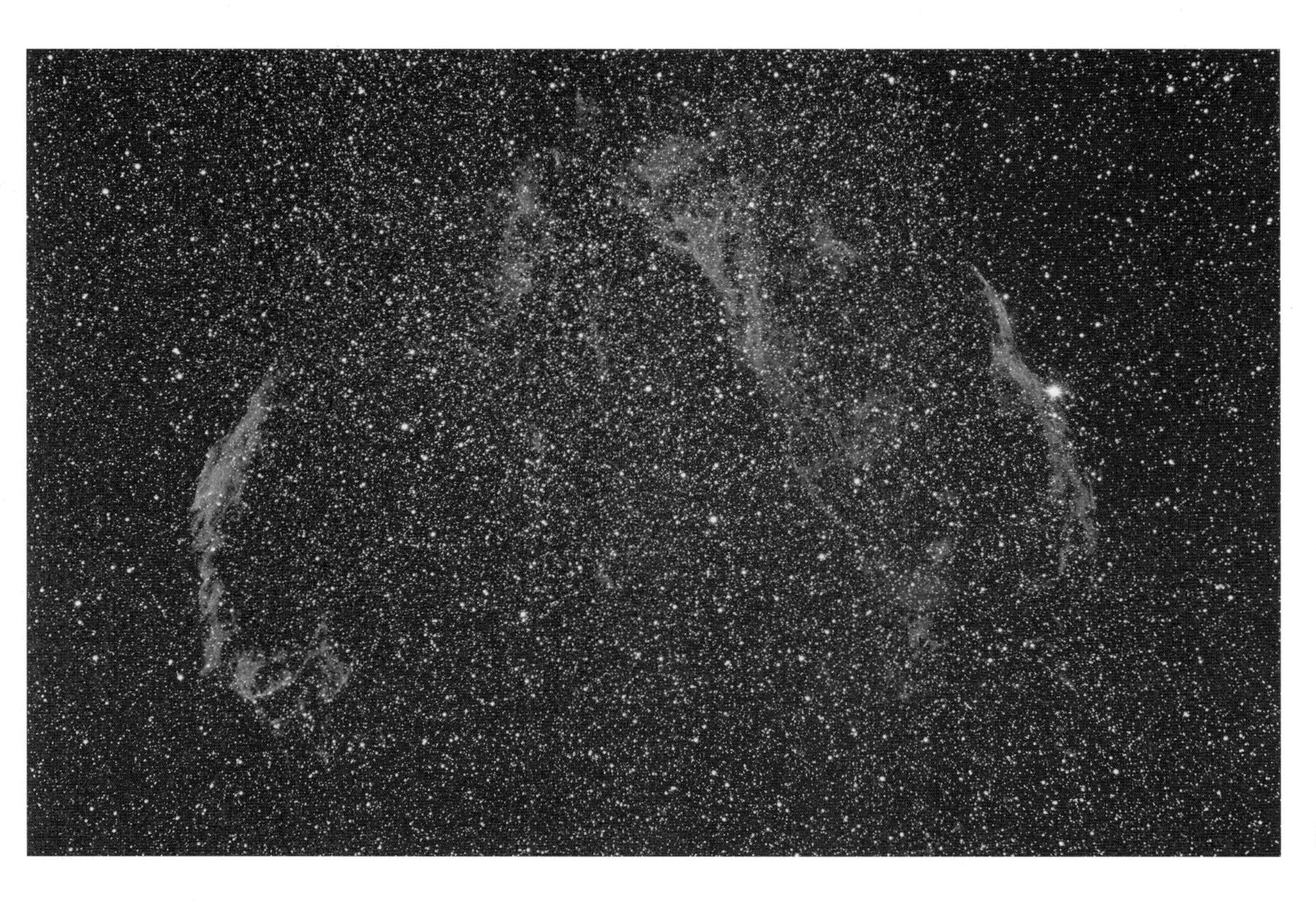

0087.jpg

Object name: IC 443 (LBN 844)
Designation: Supernova remnant (SNR) (colour)
Structures: An expanding filamentary gas cloud in Gemini with moderately strong radio source. The bright star is Eta Geminorum.
Photographic data: An Astrophysics 155 mm f/7 refractor and 35 mm camera were used with hypersenstized Kodak Pro 400 PPF film. Two 50-minute exposures were made while centred at right ascension 06 hours 32 minutes and declination +22.5 degrees. The field of view is 1.2×0.8 degrees and north is up. Digital stacking and image processing were done in the computer.

0097.jpg

Object name: M33 (NGC 598)
Designation: Galaxy (colour)
Structures: This relatively close galaxy in Triangulum is a member of the Local Group of Galaxies. Its brighter members include our Milky Way, the Magellanic Clouds, M31 and M33. It is classified as an Sc spiral and its distance from Earth is approximately 2.3 million light years. Several of its brighter emission nebulae (H-II regions) can be seen as red spots in its spiral arms.
Photographic data: An Astrophysics 155 mm f/7 refractor and Pentax 6 × 7 camera were used with unhypered Kodak Pro 400 PPF-120 film. Two 50-minute exposures were made centred at right ascension 01 hour 34 minutes and declination +30.6 degrees. The field of view is 2.8 × 1.8 degrees. M33 is 65 × 38 minutes and its brightness is 6.3 magnitude. The negatives were digitally stacked and processed in the computer.

0102.jpg

Object name: M64 (NGC 4826) – Black Eye Galaxy
Designation: Galaxy (colour)
Structures: This galaxy gets its name from the dark dust lane near the nucleus. It has a relatively high surface brightness requiring a shorter photographic exposure than most other galaxies.
Photographic data: A Celestron 11-inch f/10 SCT was used with a 35 mm camera and hypersensitized Kodak Pro 400 PPF film. A single 45-minute exposure was made while centred at right ascension 12 hours 56 minutes and declination +21.6 degrees. The galaxy's size is 10.1 × 5.4 minutes and its brightness is 9.4 magnitude. The image was processed in the computer.

Glossary of Terms

aperture the diameter of a telescope's primary mirror or objective lens
association a loose aggregation of stars e.g. OB association
asterism a group of stars not physically related which appear close together in the sky due to line of sight effects
astronomical unit the average (mean) distance between the Earth and the Sun, approximately 149,600,000 km
aurora glow in the sky, usually seen near to the Earth's geographical poles, caused by solar wind particles exciting atmospheric gas molecules
Big Bang the origin of the Universe
binary two stars in mutual orbit, held together by gravity
black hole a collapsed object which has an escape velocity greater than the speed of light
CCD (charge coupled device) a chip used for electronic imaging
cluster (star, galactic, globular, open) a group of stars bound by gravity
coma the central part of a comet, the head, as seen to the eye or through a telescope
comet small icy objects thought to originate from the outer solar system
declination one part of the co-ordinate system for locating objects on the 2-dimensional surface of the celestial sphere
disk the part of a spiral galaxy outside the nucleus which contains spiral arms
dwarf (black, red, white) a small star
eclipse an astronomical event when one body passes through the shadow of another; lunar eclipses are correctly termed, but strictly speaking a solar eclipse is an occultation
exposure the time allowed for making a photographic or electronic image
fusion nuclear reaction in which nuclei of light elements join together and release energy, a process which powers the Sun
galaxy (active, barred spiral, dwarf, elliptical, irregular, spiral) an aggregation of stars, gas and dust, one of the main building blocks of the Universe
giant (red, blue) a large star
gravity the weakest of the four forces of nature, operating between masses and important in astronomical dynamics
halo a spherical region beyond the nucleus of our Galaxy
hypersensitised film which has been treated prior to exposure to make it more sensitive to the low light levels often found in astronomical imaging
image processing manipulating a digital image by computer, e.g. to enhance contrast
infra red a region of the electromagnetic spectrum with wavelengths somewhat longer than visible light; such radiation is better able to penetrate molecular clouds
light year the distance travelled by light in 1 year, approximately 6 million million miles
Local Group a small group of galaxies, including our Milky Way and the Andromeda Galaxy, moving together through space

main sequence term describing a star undergoing first stage fusion reactions; part of a colour-magnitude diagram where such stars are located

meteor ephemeral streak of light seen in the sky when a small piece of space dust enters the Earth's atmosphere at speed and heats up rapidly; if the meteor is seen during a shower, the dust may be from a comet

nebula (absorption, dark, emission, reflection) a large volume of gas and dust in interstellar space

neutron star a collapsed, dense star which, by virtue of its mass, is able to resist further collapse into a black hole – it contains neutrons together with more exotic particles; spinning neutron stars with beamed radiation are called pulsars

nova an event in a binary system where mass transfer leads to a runaway nuclear reaction on the surface of one star, causing it to increase in brightness

nucleus (atomic, galactic) the central part of an atom or galaxy

occultation an astronomical event when one body appears to pass in front of another because of its relative motion along a line of sight

parsec the distance at which the Sun-Earth baseline extends an angle of one second of arc, approximately equal to 3.26 light years or 206,265 astronomical units

planet a relatively large body, orbiting a star, which does not emit its own visible light

planetary abbreviation for planetary nebula, a star which has expelled its outer layers of gas to leave an expanding luminous shell, and a central white dwarf star

plerion a filled supernova remnant, usually one containing a pulsar

pulsar a rotating neutron star emitting focused radiation through a beaming mechanism

quasar an active galaxy, one which has a bright nucleus dissipating vast amounts of energy

right ascension one part of the co-ordinate system for locating objects on the 2-dimensional surface of the celestial sphere

Solar System the collection of objects (planets, comets etc.) and material bound to the Sun by gravity

solar wind the stream of particles emitted, together with radiation, by the Sun

stacking a process which involves superimposing a number of negatives in the enlarger to enhance the contrast in a photographic print, or a similar process carried out in and by a computer using digital images

star a self-luminous sphere (approximately) of gas

supernova the violently explosive process involving high-mass evolved stars, as well as some stars in mass-transfer binary systems

terminator the boundary between light and darkness on a planet or its satellite

ultra violet a region of the electromagnetic spectrum with wavelengths somewhat shorter than visible light

variable a star which varies its light output due to various causes, e.g. oscillation, ionisation/recombination, dust ejection, an eclipsing binary, etc.

visible light region of the electromagnetic spectrum to which the eye is sensitive

white dwarf small hot star, the nucleus of a former main sequence star which has undergone planetary nebula ejection

X-rays short wavelength radiation emitted for example by very hot plasma such as the rarefied solar corona

year the time taken for the Earth to complete one orbit of the Sun, approximately 365.25 days

zenith that point in the sky directly above an observer's head

Subject Index

Viewing the Images on the CD-ROM

The CD-ROM with this book contains images in **JPEG** format. They are in a single directory.

You need a computer with a CD-ROM drive (of course!) and software that is able to display .JPEG images. Internet browsers such as *MS Internet Explorer* or *Netscape Navigator* can be used, but graphics packages such as *Windows-98 Imager, Paint Shop Pro* and *Adobe Photoshop* will give better results. *Paint Shop Pro* has a useful browser feature that can show all the pictures in "thumbnail" format.

You should set your monitor screen colours and resolution as high as possible for the best results, ideally at least 800 × 600 pixels, 64K colours.

The resolution of the images is such that excellent detail can be obtained if they are viewed at full screen size.

Calibrating the Images for your system

Computers, video cards, drivers, and monitors all vary in their handling of colour images. If your graphics package has a means of calibrating gamma, make sure you have set it up correctly. For final adjustment you should put image **0069.JPEG** on the screen, and compare it with the identical picture on **page 101** (in the book's Colour Plate Section). Adjust the monitor brightness and contrast controls, and if necessary your graphics package's colour, saturation, and hue controls until the pictures are as similar as you can make them. This will ensure that all the images you see are as close as possible to the originals.

Printing the Images

You can print the images with very high quality results up to A4 size (USA 10 × 8 inches). Use an inkjet printer, and set the printer drivers to "best" quality. Print on glossy paper, or even better, glossy "photo quality" paper.

Copyright